# Sustainable Development Goals Series

The **Sustainable Development Goals Series** is Springer Nature's inaugural cross-imprint book series that addresses and supports the United Nations' seventeen Sustainable Development Goals. The series fosters comprehensive research focused on these global targets and endeavours to address some of society's greatest grand challenges. The SDGs are inherently multidisciplinary, and they bring people working across different fields together and working towards a common goal. In this spirit, the Sustainable Development Goals series is the first at Springer Nature to publish books under both the Springer and Palgrave Macmillan imprints, bringing the strengths of our imprints together.

The Sustainable Development Goals Series is organized into eighteen subseries: one subseries based around each of the seventeen respective Sustainable Development Goals, and an eighteenth subseries, "Connecting the Goals", which serves as a home for volumes addressing multiple goals or studying the SDGs as a whole. Each subseries is guided by an expert Subseries Advisor with years or decades of experience studying and addressing core components of their respective Goal.

The SDG Series has a remit as broad as the SDGs themselves, and contributions are welcome from scientists, academics, policymakers, and researchers working in fields related to any of the seventeen goals. If you are interested in contributing a monograph or curated volume to the series, please contact the Publishers: Zachary Romano [Springer; zachary.romano@springer.com] and Rachael Ballard [Palgrave Macmillan; rachael.ballard@palgrave.com].

Madiha Bencekri · Seungjae Lee

# Sustainable Urban Evolution

## The 15-Minute City as a Future Paradigm

Springer

Madiha Bencekri
Department of Transportation Engineering
University of Seoul
Seoul, Korea (Republic of)

Seungjae Lee
Department of Transportation Engineering
University of Seoul
Seoul, Korea (Republic of)

ISSN 2523-3084 ISSN 2523-3092 (electronic)
Sustainable Development Goals Series
ISBN 978-981-96-5145-0 ISBN 978-981-96-5146-7 (eBook)
https://doi.org/10.1007/978-981-96-5146-7



This Springer imprint is published by the registered company Springer Nature Singapore Pte Ltd.
The registered company address is: 152 Beach Road, #21-01/04 Gateway East, Singapore 189721, Singapore

# Foreword

*Sustainable Urban Evolution: The 15-Minute City as a Future Paradigm* is a timely, comprehensive guide that moves beyond theory, offering practical strategies for urban planners, policymakers, and community leaders. It stands out by bridging theory and practice, providing real-world case studies and applications to equip professionals with the tools needed to implement the 15-minute city model. The insights shared are invaluable for cities of any size, making this book an essential resource for those committed to building sustainable and resilient urban environments.

One of the key strengths of this book is its interdisciplinary approach that integrates insights from urban and transportation planning, environmental science, and economics. This ensures a rich, adaptable framework for a 15-minute city that applies to diverse contexts worldwide. In addition, the book's emphasis on community engagement was notable. This indicates that sustainable urban transformation cannot succeed without active community involvement. This book also explores the role of smart technologies, illustrating how data-driven planning and intelligent infrastructure can enhance sustainable development. This provides a balanced analysis by addressing both the potential and challenges of integrating technology into urban planning.

Having met the authors, Prof. Seungjae Lee and Dr. Madiha Bencekri, at a conference in Seoul, we were impressed by their passion for a deep understanding of the 15-minute city. Their expertise is evident throughout this book, making it a significant contribution to the field. As an advocate of the 15-minute city concept, we believe in its transformative power.

*Sustainable Urban Evolution: The 15-Minute City as a Future Paradigm* not only furthers the conversation but also serves as a call to action. This will undoubtedly inspire professionals and communities alike to take bold steps toward creating more sustainable and inclusive cities.

Paris, France
October 2024

Carlos Moreno
Catherine Arod-Gall

# Preface

Cities worldwide face urgent challenges, from the escalating impacts of climate change to rapid urbanization and growing demand for more liveable environments. Therefore, innovative approaches to urban planning are essential. This book, *Sustainable Urban Evolution: The 15-Minute City as a Future Paradigm* explores the transformative potential of the 15-minute city concept, which reimagines urban living by ensuring that essential services and amenities are accessible within a 15-minute walk or bicycle ride. This model promotes accessibility, sustainability, inclusivity, and quality of life.

Bridging the gap between theory and practice, this book provides a comprehensive guide for transitioning urban spaces into environments that embrace the 15-minute city model. Its interdisciplinary framework draws from urban planning, environmental science, sociology, economics, and technology to present actionable strategies and holistic insights. These perspectives are interwoven to address the multifaceted challenges of contemporary cities by equipping urban planners, policymakers, and community leaders with tools for meaningful change.

Community engagement serves as a cornerstone throughout this book, highlighting the methods for fostering local participation and sharing community-led initiatives. Complementing this focus, the transformative role of smart technologies has been thoroughly explored, emphasizing data-driven decision-making and intelligent infrastructure to meet the complexities of urban evolution. Together, these themes underscore the critical importance of collaboration and innovation in shaping a 15-minute city model.

I want to begin by expressing my heartfelt gratitude to Prof. Carlos Moreno, whose groundbreaking work on the 15-minute city concept has been a profound source of inspiration for this book. My deepest thanks also go to both Profs. Moreno and Catherine Gall for their invaluable support and generosity in writing the foreword for this book. Their encouragement and belief in this work have been instrumental in its development.

Additionally, I am deeply thankful to Prof. Seungjae Lee and the University of Seoul for fostering a supportive and stimulating research environment that allowed this project to take shape. Also, I am profoundly grateful to my parents, Hassan Bencekri and Naima Naji, for their unconditional love and unwavering support.

This book is dedicated to those who envision brighter, more sustainable, resilient, and inclusive urban futures. It is both a call to action and a source of inspiration, encouraging readers to take decisive steps toward creating urban environments that are sustainable, resilient, inclusive, and liveable cities.

Seoul, Korea (Republic of)

Madiha Bencekri
Seungjae Lee

**Competing Interests** The authors declare that they have no competing interests associated with this study.

# About This Book

*Sustainable Urban Evolution: The 15-Minute City as a Future Paradigm* explores the 15-minute city concept, an urban planning framework that addresses challenges such as urban sprawl, environmental sustainability, and the need for more liveable cities. By bridging theory and practice, this book provides actionable insights, equipping urban planners, policymakers, and community leaders with practical tools for implementing this transformative model.

## Structure of the Book

This book is organized into seven (7) chapters:

- **Chapter 1: Urban Planning** lays the groundwork for exploring the historical development of urban planning and transportation systems, with an emphasis on the growing importance of sustainability in shaping modern cities.
- **Chapter 2: 15-Minute City Model** introduces the theoretical framework, benefits, and challenges of the 15-minute city, providing a clear understanding of its transformative potential.
- **Chapter 3: Time and Space Dimensions** examines the interplay of time and space in the 15-minute city model, focusing on optimizing spatial planning and temporal flexibility through innovative technologies and case studies.
- **Chapter 4: Community Engagement** highlights the essential role of community participation in the planning process, offering strategies and examples of successful community-led initiatives to foster inclusivity and equity.
- **Chapter 5: Smart 15-Minute Cities** discusses the integration of smart technologies into 15-minute city planning. Data-driven planning, intelligent infrastructure, and digital tools and applications can enhance sustainability, resilience, inclusivity, liveability, and well-being in urban environments.
- **Chapter 6: Applications of the 15-Minute City Model** focuses on implementing the model through strategies such as mobility hubs, enhanced walkability, and bikeability, while addressing social and economic concerns such as equity, inclusivity, and mental health.

- **Chapter 7: Guideline for a 15-Minute City** synthesizes key insights from previous chapters, addresses implementation challenges, and outlines a vision for adopting the 15-minute city model globally as a sustainable, inclusive, and resilient paradigm.

## Target Audience

The book caters to a diverse audience, including

- **Urban planners and city officials**: Offers actionable strategies and real-world examples to strengthen effective city planning and policymaking.
- **Community leaders and activists**: Provides tools and success stories to empower grassroots community initiatives.
- **Academics and students**: Serves as a comprehensive resource for teaching, research, and exploration of the 15-minute city model, and the urban and transportation planning.
- **Environmentalists and sustainability advocates**: Explores how the 15-minute city model advances sustainable, resilient, liveable, and inclusive urban development goals.

## Unique Selling Points

- **Innovative urban model**: Introduces the 15-minute city concept, a groundbreaking approach that enhances accessibility, fosters sustainability, and strengthens community connections.
- **Practical and actionable solutions**: Offers actionable strategies and design principles to help urban planners, policymakers, and leaders implement the 15-minute city model effectively.
- **Global case studies with local relevance**: Showcases case studies from cities worldwide, providing adaptable insights and proven practices for diverse urban contexts.
- **Empowering changemakers**: Inspires readers to become active participants in shaping sustainable cities, equipping them with tools and knowledge for real-world transformation.
- **Bridging theory with practice**: Translates complex urban planning theories into clear, actionable applications, ensuring accessibility for experts, students, and community advocates alike.

# Contents

# Abbreviations

| | |
|---|---|
| AI | Artificial Intelligence |
| BRT | Bus Rapid Transit |
| BSS | Bike Sharing System |
| CCPI | Climate Change Performance Index |
| EIA | Environmental Impact Assessment |
| GIS | Geographic Information System |
| GWR | Geographically Weighted Regression |
| IPCC | Intergovernmental Panel on Climate Change |
| LCA | Life Cycle Assessment |
| MGWR | Multiscale Geographically Weighted Regression |
| NGO | Non-Governmental Organization |
| PSS | Planning Support System |
| RES | Renewable Energy Sources |
| SDG | Sustainable Development Goals |
| SUMP | Sustainable Urban Mobility Plan |
| TOD | Transit-Oriented Development |
| UGB | Urban Growth Boundary |
| UNA | Urban Network Analysis |
| WPSS | Walkability Planning Support System |

# List of Figures

# List of Tables

# 1 Urban Planning

## 1.1 Urban Planning Evolution

Urban landscapes reflect a complex interplay of deliberate planning and organic growth, shaped over time by social, economic, and environmental forces. Understanding this evolution, from early urban theories to contemporary practices, provides valuable insights into current urban environments and helps envision sustainable cities of the future.

The history of urban planning is marked by transformative shifts. Early efforts focused on addressing industrialization's challenges, giving rise to concepts like garden cities and the modernist planning movement (Sharifi 2016). These approaches reshaped urban form and transportation, revealing their interconnected development (Zhou and Gao 2020). For instance, the rise of automobile-centric planning in the twentieth century spurred urban sprawl and fundamentally altered the design of urban spaces.

As urban dynamics evolved, planners increasingly prioritized sustainability and resilience, responding to changing demographics, economic pressures, and environmental challenges. The 15-minute city concept, championed by Carlos Moreno, embodies this progression by promoting compact, multifunctional spaces where essential services are accessible within a 15-minute walk or bike ride (Moreno et al. 2021). Cities like Paris and Melbourne have successfully adopted this model, showcasing its potential to create liveable, sustainable urban environments.

Implementing the 15-minute city, however, presents challenges. Cities with entrenched infrastructure and diverse socioeconomic landscapes face difficulties in adapting their urban and transportation systems to align with this model (Pozoukidou and Chatziyiannaki 2021). Addressing these challenges requires balancing historical urban forms with contemporary needs, rethinking urban design, and ensuring equity in access to resources and opportunities.

The 15-minute city represents a culmination of historical urban planning trends and modern aspirations. It offers a forward-thinking blueprint for cities aiming to balance sustainability, efficiency, and inclusivity, addressing the pressing challenges of urbanization in the twenty-first century.

### 1.1.1 Chronological Urban Evolution

Urban planning has undergone significant transformation, evolving from conservation-focused efforts to holistic frameworks that address environmental, social, and economic challenges. This historical progression reflects an ongoing adaptation to societal needs, technological innovations, and a deeper understanding of the interconnectedness between human and ecological systems.



M. Bencekri and S. Lee, *Sustainable Urban Evolution*, Sustainable Development Goals Series,
https://doi.org/10.1007/978-981-96-5146-7_1

#### 1.1.1.1 Early Twentieth Century: Basic Urban Planning

The formalization of urban planning began in the early twentieth century, marked by the establishment of the University of Liverpool's planning school in 1909 (Zulfiqar and Kausar 2023). During this era, Ebenezer Howard's Garden City Movement (1898) emerged as a response to industrialization's challenges, such as overcrowding and pollution. By integrating green spaces with residential, industrial, and agricultural areas, Howard laid the groundwork for sustainable urban planning (Knight 2023).

Despite its utopian ideals and critiques of isolationism, the Garden City model influenced urban planning globally, emphasizing the importance of green infrastructure and multifunctional land use (Vijay 2023; Talen 2024a). Its legacy remains visible in modern practices that prioritize balanced development and environmental harmony.

#### 1.1.1.2 1960s–1970s: Environmental Awakening

The environmental movements of the 1960s and 1970s marked a turning point in urban planning. Growing concerns about resource depletion, pollution, and urban sprawl spurred initiatives aimed at conserving natural resources and minimizing cities' ecological footprints (Ulak 2023). Landmark policies like the *Clean Air Act* (Taylor and Khan 2018) and the establishment of the Environmental Protection Agency (EPA) in the USA underscored this shift (Lee 2022). These regulations underscored the importance of environment considerations, leaving a lasting impact on urban planning (Hoel 2018).

Rachel Carson's *Silent Spring* (1962) heightened public awareness of environmental degradation (Dewantara 2024), inspiring planners to adopt innovative strategies, such as smart growth and transit-oriented development that were considered effective for managing urban expansion and reducing reliance on automobiles (Arias-Meza et al. 2023). These approaches emphasized compact, walkable communities with accessible public transportation, reducing the environmental impact of urban growth.

Planners also began integrating green infrastructure, including urban parks and green belts, to preserve biodiversity and enhance urban resilience. This era introduced frameworks like the triple bottom line, which balanced environmental, social, and economic considerations, setting the stage for modern sustainability principles.

#### 1.1.1.3 1980s–1990s: Rise of Sustainability

The environmental movements of the 1960s and the 1970s heightened the awareness of the ecological impacts of urban development, leading to the emergence of ecological urbanism (Silva et al. 2024). The 1980s and 1990s saw sustainability become a central tenet of urban planning (Qian et al. 2023), influenced by the Brundtland Commission's 1987 report defining sustainable development as meeting present needs without compromising future generations (Greve and Laustsen 2024). This era witnessed increased international cooperation, exemplified by the 1992 Earth Summit's Agenda 21, which outlined strategies for sustainable urban growth, at local, national, and global levels (Koch and Grubb 2023).

Technological advancements like Geographic Information Systems (GIS) revolutionized urban planning, enabling data-driven approaches to optimize land use and reduce ecological footprints (Batty 2013). Also, sustainability metrics, such as ecological footprints and the Human Development Index (HDI), provided tools to measure urban projects' social and environmental impacts (Long et al. 2020).

During this period, the concept of sustainable communities gained traction, focusing on cohesive, resilient, and inclusive neighbourhoods (Chen 2023). Initiatives such as mixed-use development, affordable housing programs, and enhanced public transportation systems have to ensure equitable access to resources (Benton-Short and Keeley 2023). These principles continue to influence urban planning today.

#### 1.1.1.4 Twenty-First Century: Resilient, Inclusive, and Smart

The twenty-first century has brought urgent challenges, such as climate change, rapid urbanization, and social inequities. Sustainable urban planning now prioritizes sustainability, resilience, inclusivity, equity, liveability, and technological integration (Sharma and Pareek 2024; Chen 2024).

The 15-minute city, a modern paradigm, promotes compact, multifunctional neighbourhoods where residents can access essential services within a 15-minute walk or bike ride. Cities like Paris have implemented initiatives to enhance walkability, cycling infrastructure, and local amenities, reducing reliance on cars and lowering greenhouse gas emissions (Allam et al. 2022a). Melbourne has also adopted similar strategies, highlighting the global applicability of this model (Moreno 2024). However, adapting it to diverse socioeconomic and infrastructural contexts remains a challenge.

Smart technologies, including IoT devices, big data analytics, and AI, have transformed urban planning. Innovations like smart grids, intelligent transportation systems, and real-time environmental monitoring optimize resources and improve decision-making, creating adaptive urban environments (Sharma and Pareek 2024).

Climate resilience has become a cornerstone of modern planning, with strategies such as green infrastructure, renewable energy, and climate adaptation measures protecting cities from environmental risks. Social equity is equally prioritized, with participatory planning processes ensuring inclusive urban development that benefits marginalized communities.

#### 1.1.1.5 Recapitulation

The evolution of urban planning reflects its adaptability to societal, economic, and climate challenges. From the Garden City Movement's focus on balanced development to the 15-minute city model's emphasis on accessibility and sustainability, urban planning has matured into a multidimensional discipline.

**From Conservation to Modernism**

Over the past few decades, urban planning has undergone a significant paradigm evolution (Komninos et al. 2019). From the twentieth-century focus on regionalization, social welfare, and urban renaissance, to the twenty-first expansion that includes urban morphology, participatory planning, gentrification, global cities, climate change, and political economy (Haghani et al. 2023).

The formal establishment of urban planning as a discipline began in 1909 at the University of Liverpool, but its theoretical roots date back to ancient cities like Babylon and Rome (Zulfiqar and Kausar 2023). Early urbanism focused on managing industrialization's challenges, such as overcrowding and pollution. Divergent movements, such as anarchist-inspired designs, offered alternative visions like the Garden, Radiant, and Broadacre Cities, each focusing on harmony between urban forms and social equity. . Zulfiqar and Kausar (2023) identified five stages of urban evolution (Fig. 1.1):

- Pre-history: Examining early manifestations of urban planning in ancient cities.
- Foundational years: The formalization of urban planning as a discipline in the early twentieth century.
- Modernism (rational planning): A focus on scientific, engineering-centric approaches to urban growth.
- Post-modernism: A shift towards decentralization and diversity in urban design.
- Current era: Addressing the complexities of megacities, sustainability, and smart technologies.

**Transitioning to Sustainable Frameworks**

Khavarian-Garmsir et al. (2023a, b) outlined a framework that charts urban's evolution into contemporary sustainable models (Fig. 1.2):

- Garden city: Promoting mixed-use neighbourhoods and integrated green spaces.

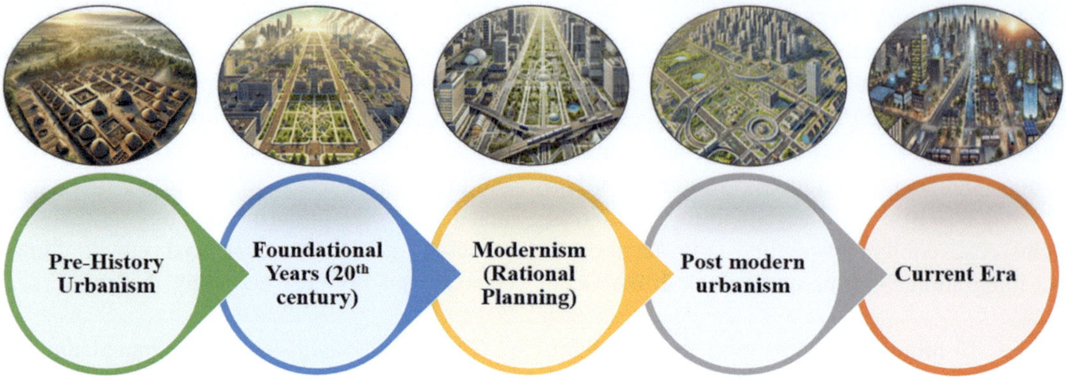


**Fig. 1.1** Five stages of urban planning evolution

**Fig. 1.2** Chronological urban planning timeline

- Neighbourhood unit: Emphasizing small, self-contained communities with local services.
- Modernist urbanism: Prioritizing large-scale, functionally segregated urban developments.
- Post-modern urbanism: Reinforcing mixed-use design and diversity.
- Eco-urbanism: Integrating environmental sustainability into urban planning.
- 15-minute city: Creating neighbourhoods that are compact, resilient, and sustainable, with daily needs accessible within a 15-minute walk or bicycle.

As urban planning evolved in the latter half of the twentieth century, it began to incorporate insights from environmental and social science, advocating for a design in harmony with natural ecosystems (Fonseca Escudero 2024). The integration of green spaces foster community interactions, cohesion, and mental well-being (Jennings et al. 2024). The enhancement of local economies attracts tourism and stimulates investment, thereby contributing to the overall vitality of cities (Zhang and Qian 2024). Technological advancements, such as remote sensing, revolutionized urban planning by enabling more data-driven and efficient decision-making processes (Batty 2013). Furthermore, incorporating equity and inclusivity into planning frameworks ensured that marginalized communities also benefit from infrastructure and sustainable development, reflecting the dynamic interplay of social, economic, and technological forces in shaping cities (Anguelovski et al. 2016).

### 1.1.2 Urban Dynamic in Shaping Cities

Urban dynamics is a multidisciplinary field that investigates the intricate processes driving the evolution, structure, and functionality of cities. It examines critical aspects such as urban morphology, land-use patterns, and the socioeconomic factors influencing them, including social structures, inequality, and economic segregation (Ramachandra et al. 2012). By integrating insights from urban planning, geography, sociology, economics, and environmental studies, urban dynamics offers a comprehensive framework to understand how urban evolution, and cities grow, adapt, and transform over time (Shaw 2021).

#### 1.1.2.1 Pillars of Urban Dynamic

To structure this analysis, Moradi and Vagnoni (2018) identified six interconnected foundational pillars of urban dynamics (Fig. 1.3):

- **Urban morphology**: The physical layout and design of cities, including street networks, building patterns, and land use.
- **Environment and sustainability**: The ecological impact of urbanization and strategies for integrating green infrastructure and sustainable practices (Giudice et al. 2023).
- **Mobility and transportation**: The networks and systems that enable movement within and between urban areas, including public transit and emerging technologies like autonomous vehicles.
- **Governance and policy**: The political and regulatory frameworks guiding urban planning, zoning, and resource allocation (Ramachandra et al. 2012).
- **Technological advancements**: The role of digital transformation, smart city technologies, and data analytics in reshaping urban services and infrastructure (Carro-Suárez et al. 2023).
- **Cultural and historical context**: The preservation of heritage, identity, and cultural diversity within modernizing environments (Alem and Domogalla 2023).

These components are deeply interconnected, dynamically influencing urban development. For example, innovations in transportation, such as electric buses, drive transit-oriented development, reshaping land-use patterns and urban morphology to support more sustainable urban forms. Governance and sustainability policies, such as incentives for renewable energy integration, not only reduce carbon footprints but also transform urban infrastructure to align with environmental goals. Similarly, while smart city technologies improve efficiency and functionality, their success depends on pairing them with inclusive policies to ensure they bridge rather than widen socioeconomic divides.

These synergies highlight the need for a holistic approach to urban planning, where advancements in one domain reinforce progress across others. Urban dynamics examines the forces shaping cities, urbanization, technology, and environmental pressures, by integrating urban morphology, sustainability, governance, and technology. It equips planners with strategies to balance growth, equity, and heritage while leveraging technology to create resilient, inclusive, and adaptable cities. This framework addresses twenty-first-century challenges, such as globalization, resilience, equity, digital transformation, and cultural identify, guiding cities towards sustainable development in a rapidly changing world.

Urban dynamics provides a comprehensive framework to address the interconnected shaping forces of cities face in the twenty-first century, including globalization, resilience, equity, digital transformation, and cultural preservation. By integrating key elements such as urban morphology, sustainability, governance, and technology, it enables planners and decision-makers to develop strategies and tools to optimize the benefits and reduce the negative consequences of the new transformative forces of cities, enabling to balance economic growth, social inclusivity, and environment and culture preservation. Urban dynamics equips cities with the tools to navigate complex global shifts, ensuring sustainable development while fostering innovation, preserving cultural

**Fig. 1.3** Pillars of urban dynamic (Moradi and Vagnoni 2018)

identity, and promoting social cohesion in an ever-changing world.

Urban dynamics provides an integrative framework for addressing the transformative forces shaping cities in the twenty-first century, including globalization, resilience, equity, digital transformation, and cultural preservation. By combining urban morphology, sustainability, governance, and technology, it empowers planners and policymakers to create strategies that harness opportunities while minimizing adverse impacts. This approach ensures a balanced focus on economic growth, social inclusivity, environmental sustainability, and the protection of cultural heritage. Through the insights offered by urban dynamics, cities can adapt to global shifts, drive innovation, preserve their unique identities, and foster cohesive communities.

#### 1.1.2.2 Cities' Shaping Forces

Cities in the twenty-first century are shaped by multiple interconnected forces, each exerting multifaceted impacts on urban environments. These forces influence economic, social, cultural, and environmental dynamics, driving the transformation of cities into more adaptive, inclusive, and sustainable spaces. Table 1.1 illustrates the multifaceted impacts of these forces.

##### Globalization and Economic Restructuring

Globalization has transformed cities into global economic powerhouses, concentrating corporate headquarters, financial services, and advanced producer activities in major urban hubs like New York, London, Paris, Casablanca, and Tokyo (Goerzen 2024). Over 50% of multinational corporations operate in just 23 global cities, illustrating the centralization of economic power.

Economic globalization reshapes urban landscapes, affecting land use, employment distribution, and social equity (Fainstein and Fainstein 1987; Tu et al. 2018). Hence, cities must develop advanced transportation and infrastructure networks to meet the evolving demands of an integrated global market (Duan et al. 2020).

**Table 1.1** Contemporain shaping forces' impacts on cities and their solutions

| Forces | Impacts | Solutions |
|---|---|---|
| Globalization | Accelerates economic growth but widens regional and socioeconomic disparities<br>Creates challenges in integrating peripheral areas into the urban core | Strengthening regional connectivity through integrated transportation systems<br>Promoting equitable policies to ensure balanced resource distribution and inclusive development |
| Equity | Inequities undermine social cohesion and limit access to urban opportunities<br>Marginalized groups face systemic exclusion from urban benefits | Adopting inclusive planning approaches, such as participatory decision-making and community engagement<br>Implementing affordable housing initiatives and inclusive zoning policies to address systemic inequalities |
| Resilience | Reduces vulnerability to climate-induced disasters and resource shortages<br>Strengthens long-term urban adaptability and sustainability | Investing in green infrastructure, renewable energy, and sustainable land use<br>Encouraging community-oriented urban design to foster social resilience |
| Digital transformation | Enables smarter, more efficient urban services but risks perpetuating digital inequities<br>Raises concerns over data privacy and cybersecurity | Bridging the digital divide through equitable access initiatives<br>Using smart technologies to optimize urban services while strengthening cybersecurity and data protection measures |
| Cultural and heritage preservation | Enhances cultural diversity and fosters a sense of community pride<br>Attracts tourism and investment, contributing to economic growth | Conserving historical landmarks and integrating heritage preservation into urban regeneration efforts<br>Promoting cultural awareness to balance modernization with historical identity |

Additionally, shifts in production methods, moving from the Fordist model of centralized manufacturing to decentralized networks of smaller, interconnected production and logistics facilities, require cities to adapt their land-use policies to support these new economic landscapes (van Egeraat and Jacobson 2005).

However, these changes exacerbate disparities between global hubs and peripheral cities. Peripheral regions often struggle to attract investment, widening economic inequalities. Decentralized production systems, while promoting flexibility, also concentrate high-skill job opportunities in select areas, leaving lower-income populations marginalized. To address these challenges, urban strategies must balance economic growth with equitable land use and inclusive economic opportunities.

### Equity and Inclusiveness

Urbanization must prioritize equity to ensure that all residents benefit from growth. Disparities in housing, health care, education, and employment remain persistent challenges, particularly in fast-growing cities (Pavel 2024). Inclusive urban planning strategies, such as participatory planning, equitable zoning, and community land trusts, empower residents to actively shape their environments (Cooper 2024).

Curitiba, Brazil, provides a successful example of equity-driven urban planning. By integrating affordable housing with sustainable public transit systems, Curitiba has created a more inclusive urban fabric that serves residents across socioeconomic strata. However, replicating such successes requires sustained political will and investment, particularly in regions where marginalized communities face systemic barriers to participation.

### Resilience

As cities confront escalating climate change impacts and social problems, resilience has become a central focus of urban planning.

Resilient cities are designed to absorb, adapt to, and recover from shocks such as natural disasters, pandemics, and economic crises (Asgari et al. 2024).

Resilient urban planning integrates adaptive and green infrastructure, sustainable land-use practices, and social resilience-building initiatives (Kapse and Jangale 2024). For example, Rotterdam exemplifies resilience with its flood-adaptive infrastructure, including water plazas that mitigate flooding while serving as public spaces. Similarly, green infrastructure like urban forests reduces heat island effects and enhances air quality, while renewable energy systems bolster energy security.

Integrating green infrastructure and renewable energy systems into urban planning strengthens both environmental and social resilience by reducing heat islands, improving air quality, and lowering greenhouse gas emissions.

Resilience strategies also require community participation, as fostering social cohesion and equitable resource distribution can strengthen a city's adaptive capacity. Policymakers should prioritize cross-sectoral collaboration, ensuring that environmental, economic, and social resilience measures are implemented cohesively.

#### Digital Transformation and Smart Cities

Digital transformation is revolutionizing urban systems, with technologies like IoT, big data analytics, and AI enabling smarter, more efficient cities (Ayesh 2024). For instance, Singapore's AI-powered traffic management system reduces congestion and emissions by dynamically optimizing traffic flows, demonstrating the potential of digital tools to enhance urban functionality.

However, these advancements also introduce significant challenges. Without equitable access, digital transformation risks widening the digital divide, leaving marginalized populations excluded from its benefits. Additionally, concerns over data privacy, cybersecurity, and ethical governance highlight the need for robust regulatory (Mahobia and Pawar 2024). Policymakers must ensure that technological innovations are inclusive, affordable, and aligned with broader urban equity goals (Bano et al. 2024).

#### Cultural and Heritage Preservation

Cultural heritage preservation plays a critical role in maintaining the identity and diversity of cities. Historical landmarks, cultural practices, and traditional urban forms enrich urban life while fostering tourism and economic development (Moser et al. 2024). Marrakech's medina revitalization demonstrates how heritage conservation can coexist with modernization, creating vibrant urban spaces that attract both residents and visitors.

Urban regeneration projects that incorporate heritage conservation can stimulate economic development and tourism, creating vibrant and attractive urban environments (Rogerson 2024). Balancing the need for modernization with heritage preservation requires careful planning and community engagement to maintain the unique character of cities while fostering sustainable growth (Boussaa and Madandola 2024).

The interconnected forces of globalization, resilience, equity, digital transformation, and cultural heritage preservation underscore the complexity of urban development in the twenty-first century. To thrive, cities must craft strategies that harmonize economic growth with environmental sustainability and social inclusivity, mainly by focusing on enhancing urban transportation.

### 1.1.2.3 Recapitulation

Urban dynamics examines how cities evolve, integrating factors like urban morphology, sustainability, governance, technology, and culture. This interdisciplinary framework provides strategies to balance economic growth, equity, and environmental sustainability in a rapidly changing world.

- Key pillars of urban dynamics:
  - Urban morphology: Optimizing urban design.
  - Sustainability: Minimalizing ecological impacts.
  - Mobility: Advancing transportation systems.
  - Governance: Guiding planners to efficient policies.
  - Technology: Leveraging smart innovations.
  - Cultural preservation: Balancing heritage and culture with modernization.
- Cities' shaping forces:
  - Globalization: Centralizes economic power but risks disparities and infrastructure efficiency.
  - Equity: Persistent inequalities require inclusive zoning, housing, and participatory planning.
  - Resilience: Climate and social issues need enhanced green infrastructure, and energy and social policies.
  - Digital transformation: Enhances urban systems but demands ethical and inclusive practices.
  - Heritage preservation: Protects identity while promoting sustainable growth.

Urban dynamics offers a comprehensive lens to address twenty-first-century challenges. By harmonizing technology, sustainability, and cultural preservation, it equips cities to thrive as adaptable, inclusive, and resilient spaces.

### 1.1.3 Sustainable Urban Planning

Urban planning has transformed significantly in recent decades, shifting from a narrow focus on resource conservation to a more holistic framework that integrates environmental, economic, and social objectives. This evolution reflects growing awareness of the complexities of urbanization, climate change, and resource depletion, along with the imperative to address systemic inequalities in urban environments.

Modern sustainable urban planning prioritizes resilience, equity, and technological innovation, enabling cities to adapt to shifting environmental, social, and economic conditions while enhancing the quality of life for all residents. At its core, sustainable urban planning aims to balance immediate human needs with long-term ecological stewardship, ensuring the sustainability of urban systems across generations (McPhearson et al. 2015; Li 2024a).

Green (2023) stated that sustainable urban development is vital to create inclusive and environmentally and economically viable cities. This study outlined the major challenges for cities, including rapid urbanization, traffic congestion and air pollution, environmental degradation, social inequality, climate change resilience, community resistance to change, housing affordability, and limited resources (Fig. 1.4). Addressing these challenges, the study suggested solutions such as compact urban design to manage growth, promoting public transit and green mobility to reduce congestion, and fostering inclusive urban policies to address inequality. Climate-adaptive strategies, participatory planning, and leveraging public–private partnerships further enhance resilience and resource efficiency. These solutions collectively ensure equitable, sustainable, and resilient urban development, benefiting all segments of society.

This section explores the key principles that underpin sustainable planning, structured around its three core dimensions: environmental resilience, economic vitality, and social equity. These dimensions form an interconnected framework, as depicted in Fig. 1.5, which highlights the overlapping priorities that guide effective and inclusive urban development.

#### 1.1.3.1 Sustainable Planning Principles

Sustainable urban planning integrates environmental sustainability, economic growth, and social equity to create cities that are inclusive, resilient, and environmentally conscious (Panagopoulos et al. 2016). Figure 1.5 illustrates the dynamic intersections between these three pillars, showcasing areas of overlap, such as sustainable natural and built environments,

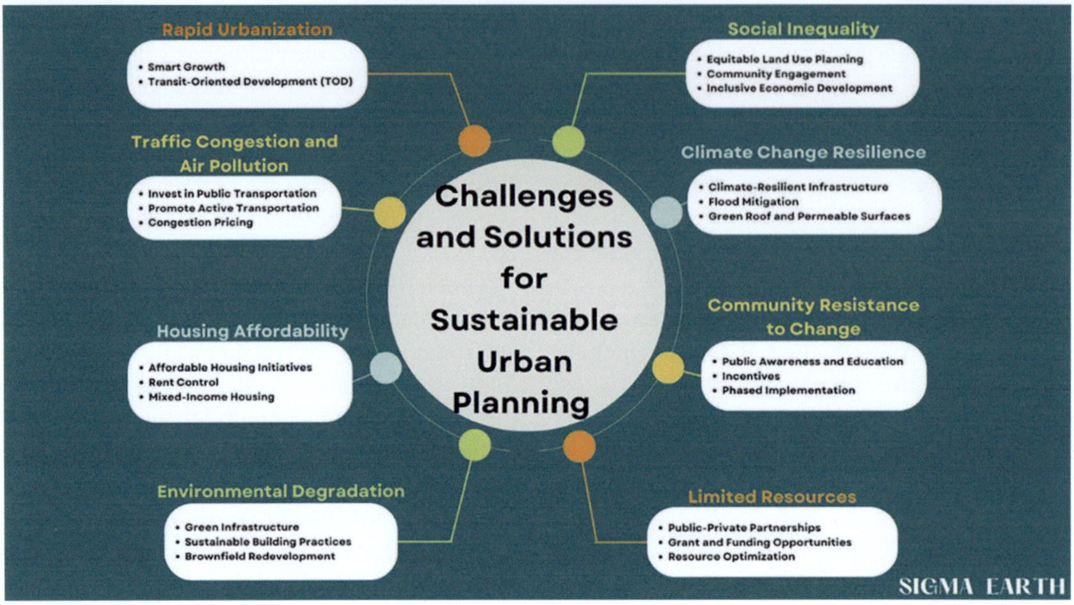


**Fig. 1.4** Sustainable planning challenges and solutions (Green 2023)

**Fig. 1.5** Sustainable urban planning principles

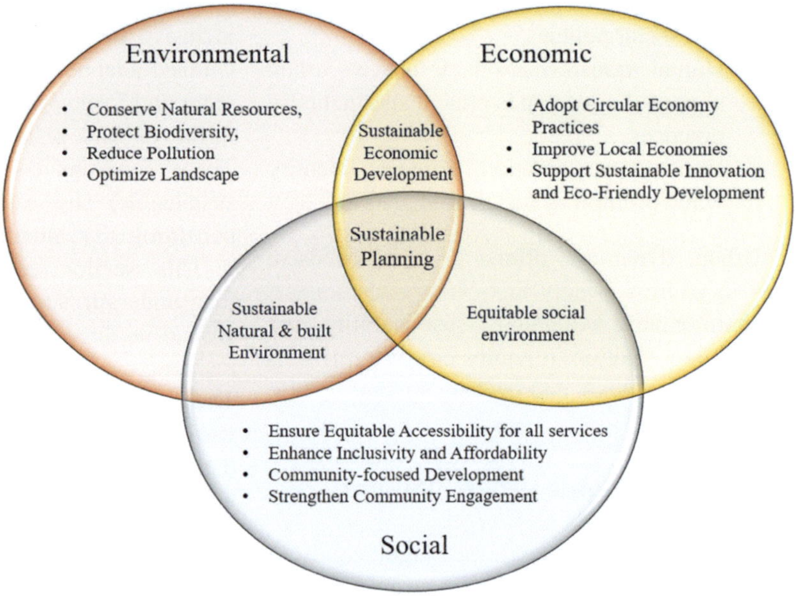


sustainable economic development, and equitable social systems.

The success of sustainable planning lies in addressing these intersections cohesively to ensure that efforts across different dimensions are mutually reinforcing. For instance, green infrastructure investments can simultaneously promote biodiversity, economic activity, and social inclusivity.

#### Environmental Dimension

The environmental dimension of sustainable urban planning emphasizes the need to conserve natural resources, protect biodiversity, mitigate climate change, and enhance urban ecosystems. These efforts ensure cities can adapt to environmental challenges while maintaining ecological balance (Prodanovic et al. 2024; Lai and Zoppi 2024).

The key strategies include:

- Green infrastructure:
  - Incorporating green roofs, urban forests, and parks into cityscapes enhances ecosystem services like carbon sequestration, air quality improvement, and temperature regulation (Adesina et al. 2024b).
  - Singapore's City in a Garden initiative exemplifies the integration of biodiversity within urban landscapes, enhancing both ecological resilience and human well-being (Srikanth and Schroepfer 2023).
  - Cities like Copenhagen have pioneered extensive cycling networks that reduce vehicle dependency and carbon emissions while encouraging healthier lifestyles (Argyros et al. 2024).
  - Urban wetlands and bioswales mitigate flooding, support biodiversity, and provide aesthetic and recreational value. These solutions blend ecological preservation with practical urban planning goals (SilwalI et al. 2023).
- Green technologies:
  - Policies including green energy-efficient and carbon reduction policies, advanced recycling systems, electric vehicles, and eco-friendly transportation, reduce emissions, and support cleaner environments (Oinam and Chakraborty 2023; Bencekri et al. 2023a).
  - Policies promoting solar, wind, and geothermal energy reduce reliance on fossil fuels while decreasing urban carbon footprints (Pandit et al. 2017; Sahu and Debsarma 2023).

These strategies intersect with the *Sustainable Natural & Built Environment*, where urban development harmonizes with ecological restoration (SilwalI et al. 2023). For instance, incorporating wetlands into urban spaces not only mitigates flooding but also fosters biodiversity, blending environmental protection with urban planning.

Sustainable urban planning must include economic and social dimensions along with the environment, to efficiently address sustainability (Loo et al. 2024). Providing equitable access to green spaces, clean energy, and efficient public transportation addresses socioeconomic disparities and fosters inclusivity (Pramanik et al. 2023). Additionally, policies supporting affordable and fair resource allocation are critical for ensuring that cities evolve into sustainable spaces (Kajiita and Kang'ethe 2024).

#### Economic Dimension

The economic dimension of sustainable urban planning seeks to balance growth with sustainability, ensuring that urban economies are resilient, diverse, and capable of adapting to disruptions such as pandemics or economic recessions (Levine et al. 2020). Integrating economic strategies with environmental conservation drives sustainable development, strengthens local economies, and encourages eco-friendly job creation (Petrie 2022).

The key strategies include:

- Circular economy practices:
  - Promoting zero-waste and shared-economy models are essential for creating sustainable urban systems (Joensuu et al. 2020).
  - Adopting approaches that help retain the value of resources while minimizing waste, through reduction, recycling systems, reuse, and resource-efficient technologies, contributing to environment efficiency (Petit-Boix and Leipold 2018).
- Efficient infrastructure development:
  - Mixed-use development, integrating residential, commercial, and industrial zones, can foster vibrant local economies, reduce commute times, and lower transportation-related emissions (Cervero and Kockelman 1997; Ewing and Cervero 2001).
  - Public transportation infrastructure enhances economic prosperity by reducing

economic disparities (Anciaes and Alhassan 2024).
  - Accessible public transportation, such as Korea's subway systems, enhances mobility, reduces economic disparities, and lowers emissions (Ahn et al. 2020).
- Support for green innovation:
  - Incentivizing renewable energy industries, green technology startups, and sustainable tourism creates eco-friendly job opportunities and stimulates economic diversification (Babic 2021).
  - Green finance is vital for energy and carbon mitigation policies, and for the broader promotion of sustainable development (Zakari 2022). For instance, tax incentives and grants are critical in promoting economy that aligns with sustainability (Svoboda 2017).
  - Leveraging technological advancements significantly enhances sustainable planning (Al-Raeei 2024). Tools such as remote sensing and big data analytics enable urban planners to make informed decisions (Adesina et al. 2024a).
  - Smart city initiatives that incorporate IoT devices and AI can optimize resource management, improve public services, and enhance energy efficiency (Zaman et al. 2024).

Sustainable urban planning integrates economic strategies with environmental conservation to create resilient, inclusive, and prosperous cities. By adopting circular economic principles, promoting economic diversification, and offering financial incentives, cities can drive economic growth while ensuring environmental and social sustainability. Additionally, community engagement, technological innovation, public awareness, and international cooperation are essential for maximizing the impact of sustainable urban planning and ensuring a higher quality of life for current and future generations.

### Social Dimension

The social dimension of sustainable urban planning prioritizes equity, inclusivity, and community well-being. This aspect ensures that urban environments provide all residents, regardless of socioeconomic status, with access to essential services, resources, and opportunities (Prentice et al. 2024).

The key strategies include:

- Inclusive urban design:
  - Models like the 15-minute city ensure that residents can access essential services within a short walk or bike ride, enhancing social cohesion and reducing car dependency (Moreno et al. 2021).
  - Implementing universal design principles ensuring equity and inclusivity for everyone, including individuals with disabilities, the elderly, and families with young children (Garofolo 2023; Müller 2023).
  - Equitable distribution of small business and retail services bridges gaps between different socioeconomic groups (Anguelovski et al. 2016).
- Community engagement and participatory planning:
  - Engaging local populations fosters support and reduces resistance to green policies (Kohli 2024).
  - Education and awareness on the benefits of green technologies, conservation efforts, and sustainable transportation can encourage more environmentally responsible behaviour (Aggarwal 2023).
  - Sharing best practices and innovative solutions helps urban planners implement effective strategies tailored to their local contexts (Li 2024a). International institutions, such as UN-Habitat and the World Bank, provide platforms for exchanges of practices and financial support (Omole et al. 2024).

Social equity and inclusion are not just important but essential components of sustainable urban planning. By actively engaging communities, applying universal design principles, educating stakeholders, addressing economic disparities, and fostering international collaboration, cities can create environments where all residents thrive.

These comprehensive efforts have led to more resilient, just, and sustainable urban communities, ensuring that everyone shares the benefits of urban development.

#### 1.1.3.2 Key Policies of Sustainable Development

Sustainable urban planning strategies aim to improve residents' well-being while ensuring cities can adapt to future challenges and thrive over the long term. It includes six (6) major components: balanced and mixed land-use and zoning, green infrastructure, social equity, economic viability, climate resilience, and technological innovation.

**Balanced Mixed Land-Use and Zoning**

Effective land-use and zoning are vital for shaping sustainable urban environments (Evans 2014). These strategies emphasize the need for optimized density, mixed-use development, and preservation of heritage and ecologically sensitive areas, including open and green spaces (Ramaiah and Avtar 2019; Raman and Roy 2019). Balanced zoning creates functional, inclusive, and environmentally conscious urban spaces, ensuring that development is sustainable both at present and in future.

A fundamental principle of sustainable urban planning is the development of mixed-use neighbourhoods, where residential, commercial, and recreational spaces coexist harmoniously. This design reduces the need for long commutes, leading to lower carbon emissions and alleviation of traffic congestion (Cervero and Kockelman 1997). Moreover, by promoting walkability and cycling, mixed-use developments encourage active lifestyles and foster vibrant community-oriented urban areas (Geyer 2024).

**Green Infrastructure**

Green infrastructure has become a cornerstone of sustainable urban planning, incorporating parks, urban forests, green roofs, and wetlands to create environmentally resilient cities (Giudice et al. 2023). These features provide numerous benefits, including biodiversity enhancement, improved air and water quality, and mitigation of urban heat islands (Tiago et al. 2024).

For instance, integrating green infrastructure such as parks, green roofs, and urban forests is crucial for enhancing biodiversity and climate change (Perrelet et al. 2024). These green spaces not only mitigate urban heat islands but also provide recreational areas that contribute to enhance residents' physical and mental well-being (Giannico et al. 2021).

Furthermore, the shift towards environmentally friendly transportation options, including public transit, biking, walking, and electric vehicles, can significantly reduce carbon emissions, ease congestion, and improve public health (Bencekri et al. 2023a). Also, integrating green transport infrastructure further enhances liveability and well-being, making cities more accessible and inclusive.

Beyond ecological advantages, green infrastructure improves human well-being by offering recreational opportunities and fostering community engagement. Integrating green infrastructure into urban plans ensures cities are not only liveable but also better equipped to withstand environmental stressors.

**Economic Viability**

Economic viability underpins resilient urban planning, emphasizing the need for local economic growth while minimizing environmental impacts. Initiatives such as supporting small businesses, fostering innovation, and promoting green industries are vital for aligning economic progress with environmental goals (Li 2024a). Sustainable economies also leverage renewable energy, circular economic principles, and efficient resource use to build long-term resilience. Urban planners must collaborate with policymakers to create incentives for sustainable investments and equitable economic growth.

**Social Equity and Inclusivity**

Social is a foundational principle of sustainable urban planning, ensuring that cities are inclusive and meet the needs of all residents, especially marginalized groups (Prentice et al. 2024).

Inclusion ensures a fair access and usage of all services and resources, bridging the social divide and creating cohesive communities. For instance, it involves ensuring that public spaces and transportation are accessible to all socioeconomic groups (Anguelovski et al. 2016).

Participatory planning models involve diverse stakeholders in decision-making, enabling urban development that reflects community needs (Anguelovski et al. 2016). Engaging the community not only enhances social sustainability but also fosters a sense of ownership and responsibility towards urban environments (Echendu 2023).

This approach enhances social cohesion and supports long-term urban sustainability. Ensuring equitable access to affordable housing, education, health care, and public transportation is vital for creating cities that cater to the diverse needs of all residents.

#### Climate Resilience

Climate resilience is a critical focus of modern urban planning (Díaz et al. 2024), as cities face increasing risks from climate change, including rising sea levels, extreme weather events, and prolonged heatwaves (Tedeschi 2024). Strategies such as integrating green infrastructure, enforcing robust building codes, integrating eco-friendly and optimal resource management, and developing comprehensive emergency response plans are essential to mitigate these risks (GÜRÇAM 2023). Resilience-focused planning not only protects vulnerable populations but also safeguards economic and environmental sustainability in the face of uncertain future challenges.

For instance, water and waste management are critical components of sustainable urban planning. Strategies such as innovative recycling, sustainable drainage systems, and water-saving techniques are vital for conserving resources and minimizing pollution (Geetha Varma 2022). Practices such as rainwater harvesting, wastewater recycling, and zero-waste initiatives are foundational to creating resilient and resource-efficient urban environments.

Achieving high energy efficiency is a cornerstone of climate resilience. It involves the promotion of renewable energy sources (García García 2022) and the adoption of energy-efficient technologies (Kiranmai et al. 2024), and developing eco-friendly infrastructure (Umoh et al. 2024). Umoh et al. (2024) highlighted that the key components of energy-efficient urban planning include building designs and retrofitting, transportation systems, and infrastructure development. For instance, retrofitting existing buildings with energy-saving technologies conserves resources and contribute to the overall sustainability of cities (Sharma et al. 2024).

Prioritizing climate resilience is essential for mitigating the impact of climate change and ensuring sustainable development of cities. Huang and Yi (2024) showed that preliminary climate adaptive policies, significantly improved urban climate adaptation and enhanced climate resilience. These adaptive policies include technological innovation, increased public participation in environmental issues, appropriate penalties for environmental damage, and lenient environmental regulation.

#### Technological Innovation

Technological innovation is integral to the advancement of sustainable urban planning (Al-Raeei 2024). The adoption of smart city technologies, including IoT devices, data analytics, and AI, offers new ways to optimize urban infrastructure and services (Zaman et al. 2024). Geographic Information Systems (GIS), big data analytics, and Internet of Things (IoT) devices allow real-time monitoring and data-driven decision-making (Batty 2013).

Smart cities present a paradigm shift in sustainable urban planning. Innovations like smart grids, predictive environmental models, and real-time monitoring enable planners to create adaptive, resource-efficient cities (Sharma and Pareek 2024). Smart policies, such as intelligent transportation systems and energy-efficient grids, reduce congestion, optimize resource allocation, and minimize environmental impacts (Nikolov 2024).

This comprehensive approach is essential for building cities that are not only sustainable, but also vibrant and equitable for all residents. However, the rapid adoption of smart technologies raises significant challenges, including concerns about data privacy, cybersecurity, and the digital divide (Gupta et al. 2020). Hence, urban planners must develop governance frameworks that ensure equitable access to smart solutions and avoid further marginalization of vulnerable populations.

#### 1.1.3.3 Recapitulation

Sustainable urban planning integrates environmental sustainability, economic vitality, and social equity to create cities that are resilient, inclusive, and adaptable to future challenges. This holistic framework addresses the growing pressures of climate change, rapid urbanization, resource depletion, and systemic inequalities by balancing human needs with long-term ecological stewardship.

Sustainable urban planning integrates three core dimensions: environmental sustainability, economic viability, and social equity. Environmental sustainability focuses on conserving resources, protecting biodiversity, and mitigating climate changes. Economic viability emphasizes circular economies, resource efficiency, and green innovation. And social equity ensures fair access to services, housing, and green spaces, with community engagement and participatory planning models. Together, these dimensions create resilient, inclusive, and sustainable cities.

Sustainable urban planning relies on six interconnected policies that address environmental, economic, and social challenges. These policies provide a comprehensive framework to create resilient, inclusive, and future-ready cities. The key policies include (Fig. 1.6):

- Balanced and mixed land use and zoning:
  - Promote compact, walkable spaces by integrating residential, commercial, and recreational spaces to reduce commuting and enhance urban vibrancy.
  - Preserve heritage sites and ecologically sensitive areas while optimizing density.
  - Example: Paris's 15-minute city model enables residents to access essential services within walking or cycling distance, fostering community cohesion and reducing emissions.
- Green infrastructure:
  - Incorporate parks, urban forests, green roofs, and wetlands into cityscapes to enhance biodiversity, mitigate heat islands, and improve air and water quality.
  - Support sustainable transportation systems like cycling networks and eco-friendly public transit.
  - Examples: Singapore's *City in a Garden* demonstrates ecological and social benefits, while Copenhagen's cycling infrastructure reduces car dependency and emissions.
- Economic viability:
  - Align growth with sustainability by fostering circular economies, resource efficiency, and green industries.
  - Invest in public transit and renewable energy to create eco-friendly jobs and reduce emissions.
  - Incentivize sustainable practices through tax breaks, grants, and green finance.
  - Examples: Korea's subway network enhances mobility and economic equity, while Germany's retrofitting initiatives reduce energy costs and carbon footprints.
- Social equity and inclusivity:
  - Ensure equitable access to essential services, affordable housing, green spaces, and public transportation for all residents, especially marginalized groups.
  - Promote participatory planning through public consultations, budgeting models, and educational initiatives to empower communities.
  - Apply universal design principles to guarantee accessibility for individuals with disabilities, seniors, and low-income populations.
  - Example: Porto Alegre's participatory budgeting model fosters trust and inclusivity by involving residents in urban decision-making.

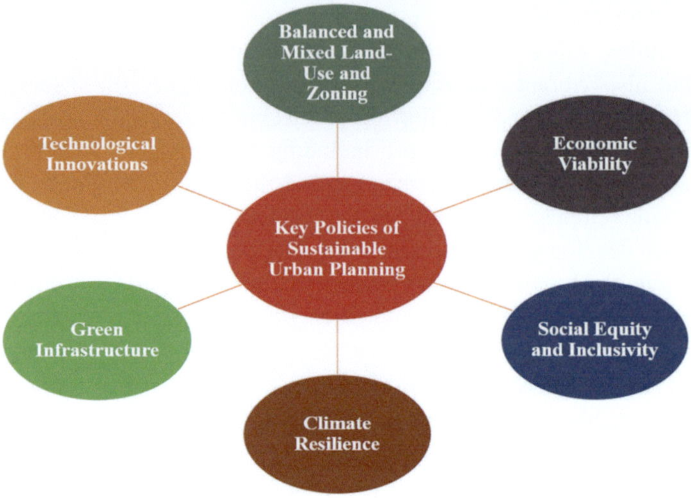


**Fig. 1.6** Key policies for sustainable urban planning

- Climate resilience:
  - Address climate risks such as extreme weather events and resource scarcity by integrating green infrastructure and enforcing energy-efficient building codes.
  - Implement innovative water and waste management strategies, including rainwater harvesting, wastewater recycling, and zero-waste initiatives.
  - Examples: Cape Town's water conservation efforts and Germany's energy retrofitting programs demonstrate proactive resource management and resilience-building.
- Technological innovation:
  - Leverage tools like IoT, AI, GIS, and big data to optimize urban systems, improve resource management, and enhance urban services.
  - Develop smart policies for energy-efficient grids, intelligent transportation systems, and predictive environmental models.
  - Address challenges such as data privacy, cybersecurity, and equitable access to ensure technology benefits all residents.
  - Examples: Real-time monitoring systems and smart grids enable adaptive and resource-efficient cities, with enhanced energy efficiency, optimal urban design, and reduced environmental impacts.

These six policies, balanced land use and zoning, green infrastructure, economic viability, social equity, climate resilience, and technological innovation, offer a comprehensive framework for sustainable urban planning. They work in tandem to create cities that are resilient, inclusive, and environmentally sound.

The interplay between these policies is vital: compact land use supports sustainable transportation; green infrastructure bolsters climate resilience; and participatory planning ensures equitable access to these benefits. Urban planners, policymakers, and communities must collaborate to implement these policies effectively, ensuring vibrant, sustainable cities for future generations.

## 1.2 Urban Form and Transportation

Urban form, encompassing the physical and non-physical characteristics of cities, evolves through a complex interplay of factors such as building design, spatial layout, and societal dynamics. As (Oliveira 2016) notes, urban spaces function hierarchically, with individual elements collectively shaping entire systems. Similarly, Williams (2014) describes cities as dynamic

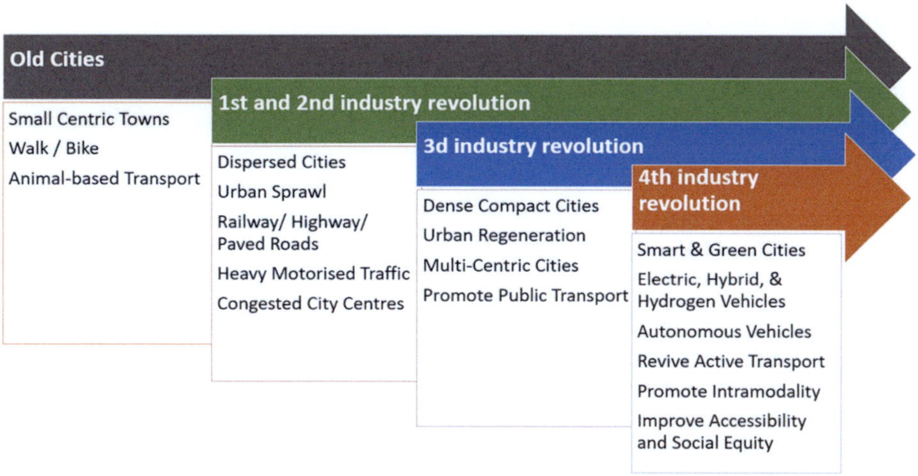


**Fig. 1.7** Evolution of urban form and transportation (Lee and Bencekri 2021)

systems where slower-changing urban forms interact with dynamic flows of capital, people, and culture. Rode et al. (2017) stated that transportation influencing various aspects such as congestion, emissions, accessibility, safety, and vehicle ownership. While, the *American NACTO Urban Street Design Guide* underscores the multifunctional role of streets, emphasizing their importance in functionality, movement, and safety (NACTO 2012).

Understanding the historical evolution of the relationship between urban form and transportation is vital for addressing contemporary challenges and creating sustainable, inclusive, and liveable cities.

## 1.2.1 Evolution Background

The historical evolution of urban forms and transportation systems, as illustrated in Fig. 1.7, reflects the interplay between technological advancements and societal needs. This progression can be categorized into four distinct phases, each defined by a transformative industrial revolution.

### 1.2.1.1 Old Cities: Pre-industrial Era

Before the industrial era, cities were small, compact, and centred around human-scale activities. Mobility primarily relied on walking, cycling, and animal-based transportation. Streets served multifunctional purposes, acting as both thoroughfares and vibrant spaces for economic, social, and cultural exchange (Marshall 2004). These early cities were compact with high accessibility and proximity.

### 1.2.1.2 1st and 2nd Industrial Revolutions: Urban Sprawl

The 1st and 2nd industrial revolutions introduced groundbreaking technologies like steam engines, railways, and automobiles, transforming urban structures. Railways improved inter-city connectivity, while highways facilitated suburban expansion. These developments shifted cities from compact, pedestrian-oriented layouts to more dispersed forms, characterized by urban sprawl and reliance on motorized vehicles.

The emergence of railways and paved roads reshaped city layouts, allowing for larger metropolitan regions but contributing to congested city centres and the decline of traditional street life. Suburbanization grew alongside the automobile industry, creating quieter but less dynamic neighbourhoods. This era prioritized functionality and economic growth but often overlooked social equity and environmental sustainability (Jabareen 2006).

### 1.2.1.3 3rd Industrial Revolution: Compact Cities

The 3rd industrial revolution introduced urban regeneration and multi-centric city structures (Lee and Bencekri 2021). This phase focused on addressing the challenges of sprawl and congestion by promoting dense, compact cities, and revitalizing urban cores (Ewing and Cervero 2010).

Public transportation systems, such as subways and light rail networks, gained prominence as cities sought to reduce car dependency and improve mobility. However, the decrease of reliance on car is very modest (Stevens 2017), whereas the over-densification increased travel distances for non-work trips (Burton 2000).

Urban planning shifted towards multi-centric layouts, where multiple hubs within a city provided localized economic and social opportunities. Promoting public transportation became a critical strategy during this phase, reflecting an increasing awareness of environmental impacts and the inefficiency of car-dominated systems. This decentralization improved access to jobs and services while reducing pressure on central business districts.

### 1.2.1.4 4th Industrial Revolution: Smart and Green Cities

The 4th industrial revolution represents a transformative era driven by digitalization, sustainability, and inclusivity (Lee and Bencekri 2021). Urban areas are now embracing smart and green models that prioritize equity, resilience, and technological innovation.

Sustainable transportation planning is focusing on to revive active transport modes such as walking and cycling, which had declined in earlier industrial phases (Pucher et al. 2011). Also, it depends strongly on promoting public transit, by enhancing safety, convenience, cost, and comfort these aspects (Schneider 2013). Additionally, (El-Geneidy et al. 2014) and (Rychlewski 2016) advocated for shorter walkable distances to increase user public transit attraction.

Furthermore, transportation systems are being revolutionized through the adoption of electric, hybrid, and hydrogen vehicles as well as autonomous vehicles. These technologies aim to reduce emissions, improve energy efficiency, and enhance urban mobility.

Cities are also increasingly embracing intermodality, designing integrated transportation networks that seamlessly connect various modes of travel, such as buses, trains, and bike-sharing systems. This integration enhances accessibility while reducing dependence on private cars.

Beyond technological advancements, the 4th industrial revolution prioritizes social equity, ensuring that urban systems are inclusive and accessible to all residents. Planners are focusing on reducing spatial inequalities and improving access to public services, particularly for marginalized communities.

### 1.2.1.5 Key Transportation Challenges

Urban transportation systems face numerous interrelated challenges that adversely affect urban environments, economies, and societies. These issues, summarized in Fig. 1.8, encompass congestion, environmental degradation, safety concerns, social inequities, and the lack of integrated mobility solutions. Addressing these challenges demands a holistic approach to urban planning that prioritizes sustainability, equity, and safety.

Congestion remains one of the most visible and detrimental challenges in urban mobility. The increasing number of vehicles competing for limited road space results in time delays, which translate into decreased economic productivity and higher operational costs for businesses (Petrov et al. 2019).

Beyond economic impacts, congestion is a significant contributor to air pollution due to the idling of vehicles, which emit harmful pollutants such as nitrogen oxides and particulate matter. This deteriorates air quality and poses severe health risks, including respiratory diseases and cardiovascular problems.

Congestion also exacerbates mental health problems, as prolonged commuting times contribute to stress and reduced quality of life. To address these cascading effects, urban planners

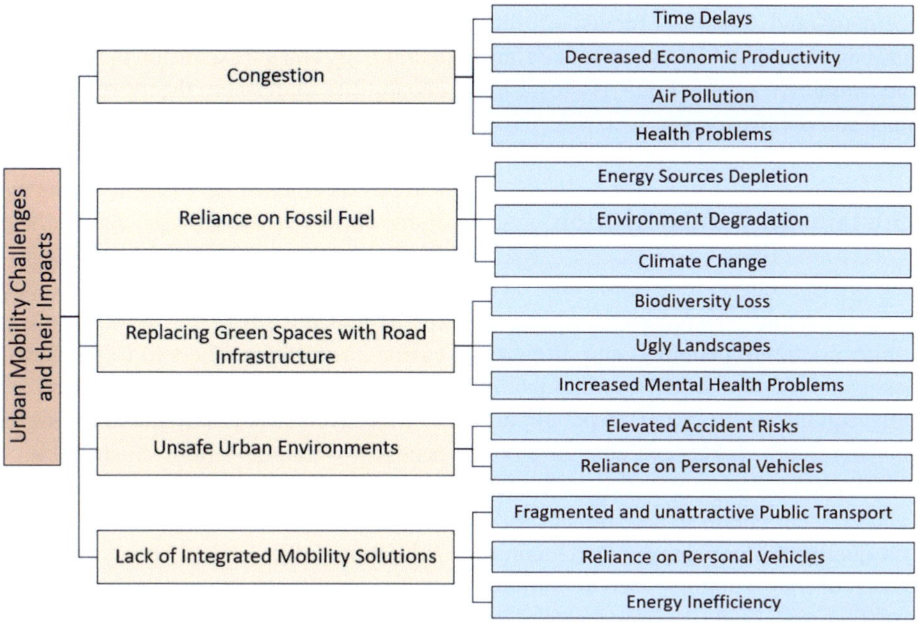


**Fig. 1.8** Urban transportation challenges

must prioritize efficient public transportation systems, discourage private car dependency, and implement smart traffic management solutions.

Urban transportation's heavy reliance on fossil fuels accelerates energy depletion, climate change, and environmental degradation (Haines 2000). The burning of fossil fuels for transportation is a leading cause of greenhouse gas emissions, which contribute to global warming, while the spatial allocation of road infrastructure often comes at the expense of green spaces (Geneletti 2003), reducing biodiversity and affecting public health (Southworth 2016; Southon et al. 2018).

Efforts to reduce environmental impacts must focus on promoting green infrastructure, adopting renewable energy-powered transportation, and integrating nature into urban designs to restore biodiversity and enhance urban resilience.

Urban mobility often perpetuates social inequalities. Low-income communities, typically located in areas with poor transit connectivity, face barriers to accessing employment, education, and essential services (Mirzaee and Wang 2020). This lack of equitable access reinforces cycles of poverty and marginalization.

Safety is a critical but often underemphasized challenge in urban mobility. Traffic accidents and pedestrian fatalities remain alarmingly high in cities where urban planning prioritizes vehicles over pedestrians and cyclists (Dumbaugh and Li 2010).

The lack of safe infrastructure discourages active transportation modes, such as walking and cycling, leading to higher dependence on cars and contributing to sedentary lifestyles. This has significant public health implications, increasing the prevalence of obesity, cardiovascular diseases, and other health issues (Congiu et al. 2019).

Reliance on private vehicles not only creates economic burdens for low-income families, who must allocate significant resources to transportation costs, but also exacerbates safety concerns due to inadequate infrastructure for pedestrians and cyclists. To address these interconnected challenges, policymakers and urban planners must invest in affordable and efficient public transportation, improve connectivity in underserved areas, and design pedestrian-friendly environments that prioritize protected bike lanes and enhance enforcement of traffic regulations to reduce accidents and fatalities. Inclusive urban design that ensures equitable access to safe and sustainable mobility is essential for fostering social equity and public safety.

### 1.2.2 Sustainable Transportation

Sustainable transportation transcends the mere adoption of greener modes; it represents a comprehensive shift in urban planning and lifestyle, encompassing the environmentally, economically, and socially equitable movement of people and goods (Banister 2008). It is about changing how cities are built and how to provide fair movement services and efficient public spaces and design to enhance the quality of life (Fejza 2022). It includes diverse modes of transportation, such as walking, cycling, public transit, and eco-friendly vehicles, each of which plays a vital role in minimizing the ecological footprint of transportation, reducing resource consumption, and ensuring equitable access for all within the community.

By implementing sustainable mobility practices, emissions from the transportation sector can be significantly reduced, leading to improved air quality and health outcomes (Maizlish et al. 2013). Such initiatives not only contribute to economic growth and social equity but also enhance the overall urban living experience. For instance, investment in robust public transportation networks broadens access to employment, education, and health care, offering a lifeline to lower-income brackets (Bocarejo and Urrego 2022). Similarly, encouraging active transportation modes, such as walking and cycling, not only eases congestion but also promotes healthier, more engaged communities (Stroope 2023).

Decarbonization is essential for advancing sustainable mobility by significantly reducing the environmental impact of transportation. A key strategy is the adoption of electric vehicles (EVs) and the development of renewable energy-powered charging infrastructure (Gil-García et al. 2021), which promise a drastic reduction in emissions compared to traditional fossil fuel vehicles. This shift is vital for meeting SDG 13: Climate Action Goals and mitigating the adverse effects of climate change. Moreover, innovations in battery technology and the expansion of renewable energy capabilities underscore the move towards greener transportation options (Xiao 2024). The success of decarbonization relies not only on technological advancements, but also on supportive policy frameworks that incentivize sustainable practices (E. Ezenwa et al. 2024). For instance, prioritizing and enhancing public transportation and cycling to encourage a shift from private car use, showcasing a model for others to follow, and aligning with broader environmental and social objectives.

Moreover, it is important to ensure equitable access to transportation. Sustainable mobility seeks to dismantle the economic and social barriers faced by marginalized communities. Addressing affordability, accessibility, and inclusivity issues head-on, sustainable mobility initiatives aim to forge a transportation system that serves the needs of all societal segments (Litman 2022). Specifically, the creation of comprehensive public transit networks that connect underserved areas to critical services and job opportunities is crucial.

Similarly, the promotion of micro-mobility solutions such as accessible bike-sharing programs and pedestrian-friendly pathways ensures that diverse community needs are met (Alka et al. 2024). Creating well-connected public transit systems and fostering micro-mobility solutions cater to diverse needs, thereby advancing SDG 10 (Reduced Inequalities). These measures not only facilitate greater social inclusion but also contribute to a reduction in congestion and pollution, aligning with environmental sustainability objectives.

The 15-minute city concept can reduce carbon emissions and enhance public health by encouraging active modes of transportation (Moreno et al. 2021). Furthermore, it supports the growth of local economies and advances social equity by making daily necessities readily accessible to all community members. By prioritizing mixed-use development, optimizing public transportation systems, and ensuring seamless first-last mile connectivity, this concept aligns with the principles of sustainable mobility and SDG-11 goals related to sustainable cities and communities.

**Fig. 1.9** Comprehensive framework for the 15-minute city

It envisions a future in which urban areas are designed to be more walkable and bike-friendly, thus becoming universally accessible and significantly improving urban liveability.

### 1.2.3 Transportation Within a 15-Minute City

In response to these pressing challenges, the 15-minute city concept has gained global attention as an integrated solution to the challenges of urban mobility, including traffic congestion, pollution, and social inequities.

The 15-minute city model exemplifies this transformation by integrating advanced technologies, innovative design strategies, and participatory planning to create accessible, inclusive, equitable, and sustainable urban spaces (Allam et al. 2024). Shoina et al. (2024) linked the success of this model to the integration of a holistic policy framework, accessibility, sustainability, community engagement, and economic resilience, as depicted in Fig. 1.9.

***Policy frameworks*** form the backbone of the 15-minute city model. These policies guide cities in creating cohesive and inclusive environments that meet the needs of all residents while addressing key urban challenges, forming the backbone of the 15-minute city model (Shoina et al. 2024). For instance, Paris's redesign of neighbourhoods to include mixed-use developments showcases how policies can improve accessibility and reduce inequalities (Lee 2024). Therefore, tailored policies are essential to accommodate diverse urban contexts.

***Accessibility*** is at the heart of the 15-minute city model (Moreno et al. 2021). Policies must promote compact urban design that reduces travel times, supports walkability, and ensures proximity to essential services. Public transit and walking-friendly infrastructure are critical elements that align accessibility with sustainability goals (Yannis and Chaziris 2022).

***Sustainability*** is another pillar of the 15-minute city, driven by policies that prioritize active models, green infrastructure, and energy efficiency. Urban parks, rooftop gardens, and renewable energy systems contribute to environmental resilience while improving air quality, biodiversity, and public health. Singapore's *City in a Garden* initiative exemplifies how policies can embed sustainability into urban design (Yeo 2019).

Policies that encourage ***community participatory planning***, such as public forums, neighbourhood councils, and inclusive surveys, ensure that development reflects the needs and aspirations of local populations (Newell et al. 2020). By involving marginalized groups, these policies promote equity, inclusivity, and social cohesion (Whitney and Ledsham 2024).

The 15-minute city also offers ***economic resilience***, for instance, by stimulating local businesses, improving land value, and creating of jobs (Yang 2023). Policymakers must integrate economic growth with environmental sustainability, and collaborate with stakeholders to create economic incentives, such as grants for small businesses or tax breaks for sustainable enterprises, ensuring that economic growth aligns with sustainability goals (Wang et al. 2014). The 15-minute city model holds immense promises, but its implementation requires careful consideration of serious challenges.

In conclusion, the 15-minute city model provides a transformative vision for urban development, addressing critical challenges such as traffic congestion, environmental degradation, and social inequities through a holistic framework that emphasizes accessibility, sustainability, community engagement, and economic resilience. By integrating tailored policies, advanced technologies, and participatory planning processes, this model fosters vibrant, inclusive, and self-sufficient communities. However, realizing the full potential of the 15-minute city requires a concerted effort from policymakers, urban planners, businesses, and residents, as well as a commitment to overcoming structural and financial barriers. As cities around the world continue to adopt and adapt this framework, the 15-minute city serves as a powerful blueprint for building more equitable, resilient, and sustainable urban futures.

## 1.3 Book Significance

Globally, escalating levels of $CO_2$ emissions present a critical challenge, necessitating immediate and radical action to achieve the ambitious goal of achieving carbon neutrality by 2050. The urgency of the situation is further accentuated by the Climate Change Performance Index (CCPI), a tool that assesses the effectiveness of national climate policies (Fig. 1.10). The CCPI's rankings reveal that many countries are falling behind in their efforts to combat climate change, underscoring a global shortfall in effectively addressing this existential threat.

Achieving the target of carbon neutrality by 2050 is not only an environmental imperative but also a necessity for sustainable human existence. It demands a shift towards sustainable energy practices, adoption of innovative policies and technological solutions, and adoption of new urban policies. This transition opens the door to innovative urban planning concepts, such as a 15-minute city. This model envisions cities where residents can access their needs within a 15-minute walk or bike ride, thereby promoting sustainability, community, and local economies.

The 15-minute city concept is a vital urban planning model aimed at creating spaces where essential services are accessible within a 15-minute walk or bike ride, thereby reducing the need for car trips and shortening transit times (Moreno et al. 2021). This model can significantly aid in achieving decarbonization goals and fostering sustainable, accessible urban environments. However, its successful implementation requires extensive planning and public participation.

While a 15-minute city offers environmental, social, and economic benefits, actualizing it involves complexities. Key obstacles include integrating essential factors, such as land use, participatory governance, and connectivity, into traditional planning methods, which often overlook these aspects (Geertman and Stillwell 2009). Additionally, existing methods face limitations in scenario building, impacting their ability to provide comprehensive analyses of various urban development possibilities (Batty 2013). This study addresses these challenges by providing evidence-based insights and tools to help urban planners and policymakers develop effective 15-minute cities. It seeks to bridge the gap between theoretical understanding and practical applications, creating a tangible roadmap for establishing sustainable and equitable cities. This study aims to enhance the integrated understanding of urban and transportation systems, contribute to the broader

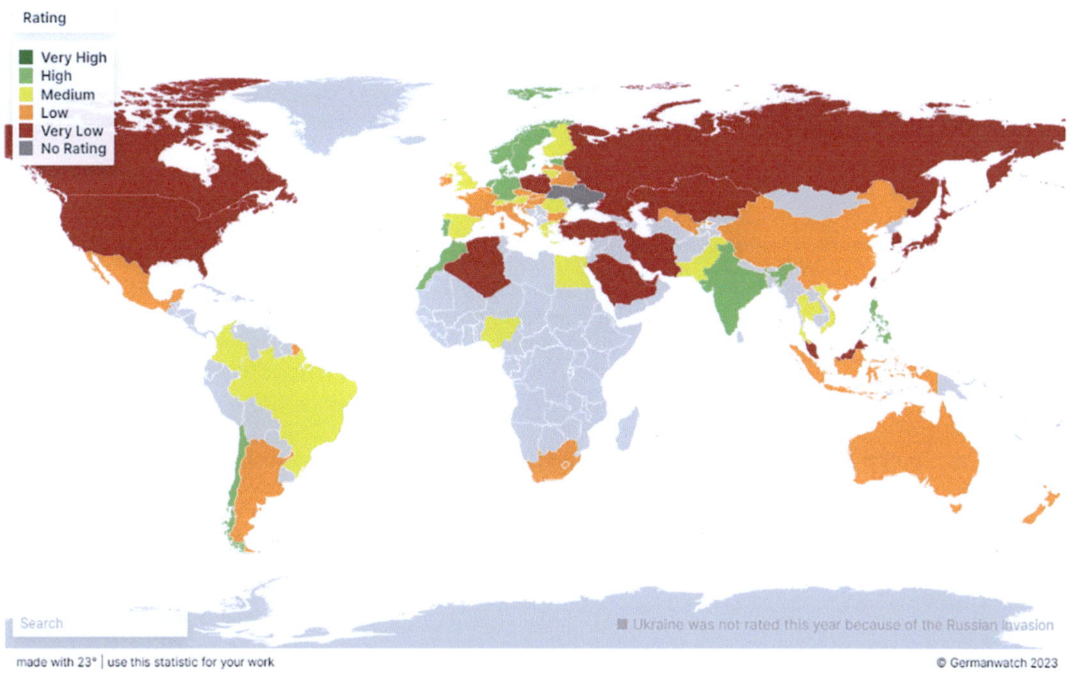


**Fig. 1.10** World CCPI index ranking (*Source* Germanwatch, 2023)

goals of sustainable urban development, and serve as a resource for researchers, practitioners, and policymakers.

Within the realm of the discourse surrounding the 15-minute city concept, a multitude of scholarly works have emerged, each contributing unique perspectives and insights into this urban planning paradigm. To fully appreciate the significance of the book "Sustainable Urban Evolution: The 15-minute city as a Future Paradigm," it is crucial to contextualize its contributions within the broader landscape of literature on this subject (Table 1.2).

One of the seminal works in this domain is "*The 15 Minute City: Global Change Through Local Living*" by Whittle(2021) to serve as an in-depth interrogation of the 15-minute city concept, meticulously elucidating its merits, demerits, and potential to revolutionize contemporary urban living. The "*15-minute city*" by Cee (2023) provides an introductory overview of the 15-minute city concept and its potential benefits. The "*15 Minute City in 10 Minutes: Diving into what they are, quickly*" by Giuseppe Giuseppe Melillo (2023) offers a concise overview of the 15-minute city concept. Bob Lilli-Drake's "*15 Minute Cities*" (2023) discusses the potential of 15-minute cities to enhance the quality of life and reduce dependence on automobiles (Lilli-Drake 2023).

Moreno et al. (2024)'s "*The 15-minute city*" introduces a transformative urban planning model that ensures essential services are within a short walk or bike ride from home. Often hailed as the "godfather" of this concept, Moreno's work provides strategies for reducing car dependency and enhancing urban sustainability. His book offers actionable insights for policymakers and urban planners by presenting research-backed solutions to long-standing urban challenges. This essential resource is invaluable for those seeking modern human-centric approaches to city planning.

Webb (2023)'s "*15 Minute City Debating its Necessity*" (2023) examines the necessity and functionality of the 15-minute city and offers an extensive exploration of the pros and cons, integrating diverse perspectives. In contrast, "*The 15-minute city: A Utopian Dream or a Dystopian*

**Table 1.2** Existing competitive books

| Book | Author(s) | Year | Key contributions | Comparison with this book |
|---|---|---|---|---|
| The 15 Minute City: Global Change Through Local Living | Natalie Whittle | 2021 | Interrogates the 15-minute city concept: its pros, cons, and potential to revolutionize modern living | This book takes a deeper and interdisciplinary dive into the 15-minute city model, emphasizing community engagement, sustainable practices, and urban mobility solutions |
| The 15-Minute City: A Utopian Dream or a Dystopian Nightmare? | Rene Mulberry | 2023 | Explores both the potential benefits and drawbacks of the 15-minute city concept | While Rene Mulberry's work critically examines the 15-minute city, "The Dynamic City" offers a comprehensive study and practical insights for implementation |
| The Dangers of 15 Minute Cities: An Undemocratic and Violation of Human Rights | Dolain Mwamba | 2023 | Raises concerns about the potential for 15-minute cities to infringe on individual rights and limit personal freedoms | This book contrasts Mwamba's concerns by outlining strategies for inclusive and democratic urban planning that respects individual rights and ensure equity and inclusivity for all different demographic and socioeconomic groups |
| 15-Minute City | Keith Cee | 2023 | Provides an overview of the 15-minute city concept and its potential benefits | Expanding on Keith Cee's introductory perspective, this book relies on real case studies and actionable strategies for urban transformation under the 15-minute city concept |
| 15 Minute Cities | Bob Lilli-Drake | 2023 | Discusses the potential of 15-minute cities to improve quality of life and reduce reliance on cars | Building on Lilli-Drake's positive outlook, this book provides a more detailed analysis of urban design and its impact on community well-being |
| 15 Minute City Debating its Necessity | Hugh Webb | 2023 | Examines the pros and cons of the 15-minute city concept | This book provides a broader narrative that integrates various perspectives on the necessity and functionality of the 15-minute city |
| 15 Minute City in 10 Minutes: Diving into What They Are, Quickly | Giuseppe Melillo | 2023 | Provides a quick overview of the 15-minute city concept | This goes beyond a quick overview to offer an in-depth guide on the principles, implementation, and future of 15-minute cities |

*Nightmare?*" authored by MULBERY (2023) adopted a more critical stance, offering a comprehensive examination of both the potential benefits and drawbacks associated with the 15-minute city concept. Also, Mwamba (2023)'s "*The Dangers of 15 Minute Cities: An Undemocratic and Violation of Human Rights*" raises valid concerns regarding the potential for 15-minute cities to infringe on individual rights and freedoms.

In summary, the book "Sustainable Urban Evolution: The 15-minute city as a Future Paradigm" distinguishes itself through a comprehensive and balanced approach to the 15-minute city concept, emphasizing community engagement, interdisciplinary research, and an actionable framework. Its unique contributions are significant in addressing gaps in the literature and making it a valuable resource for diverse audiences, including scholars, urban planners, and policymakers.

A notable contribution of this book is its emphasis on community engagement. While previous works have explored the theoretical aspects of the 15-minute city, this book goes further by highlighting the pivotal role of communities in its realization. This underscores the importance of actively involving residents in planning and decision-making processes, fostering a sense of ownership and commitment to sustainable urban evolution. In doing so, the book bridges the gap between theory and practice, offering practical insights that are crucial for urban planners and policymakers.

Additionally, this book adopts an interdisciplinary research approach, setting it apart from many prior works that often focus on singular aspects of a 15-minute city. It recognizes the complex interplay of various factors in urban development and integrates insights from multiple disciplines, including urban planning, sociology, environmental science, and economics. This interdisciplinary perspective enriches discourse by acknowledging the interconnectedness of urban systems and the need for holistic solutions.

Furthermore, the book provides an actionable framework for implementing the 15-minute city concept, a feature that distinguishes it from other works that primarily offer critical analyses or theoretical discussions. It offers tangible strategies, case studies, and real-world examples that can be applied directly by urban planners, policymakers, and researchers. This practical guidance enhances the book's utility and relevance in shaping a sustainable urban evolution.

While the existing literature has provided valuable insights, this book provides a comprehensive perspective that integrates community engagement, interdisciplinary research, and actionable guidance. These contributions are vital in the context of urban planning and sustainable development because they address real-world challenges and empower stakeholders with the knowledge and tools needed to usher in a new era of sustainable urban evolution.

## References


Adesina AA, Okwandu AC, Nwokediegwu ZQS (2024a) Geo-information systems in urban planning: a review of business and environmental implications. Magna Sci Adv Res Rev 11:352–367. https://doi.org/10.30574/msarr.2024.11.1.0100

Adesina AA, Okwandu AC, Nwokediegwu ZQS (2024b) Towards sustainable urban development: conceptualizing green infrastructure and its impact on urban planning. Int J Appl Res Soc Sci 6:1274–1296. https://doi.org/10.51594/ijarss.v6i7.1266

Aggarwal DD (2023) Green education for a sustainable future. J Environ Impact Manag Policy 27–30. https://doi.org/10.55529/jeimp.34.27.30

Ahn K, Jang H, Song Y (2020) Economic impacts of being close to subway networks: a case study of Korean metropolitan areas. Res Transp Econ 83:100900. https://doi.org/10.1016/j.retrec.2020.100900

Alem G, Domogalla F (2023) The alley as an expression of urban form: Understanding the logic of African urban dynamics. J Urban Aff 45:461–482. https://doi.org/10.1080/07352166.2022.2070496

Alka TA, Sreenivasan A, Suresh M (2024) Wheel of change: a systematic literature review on innovation and entrepreneurship in micro mobility solutions. Transp Econ Manag 2:154–168. https://doi.org/10.1016/j.team.2024.06.004

Allam Z, Bibri SE, Chabaud D, Moreno C (2022a) The theoretical, practical, and technological foundations of the 15-minute city model: proximity and its environmental, social and economic benefits for sustainability. Energies 15:6042. https://doi.org/10.3390/en15166042

Allam Z, Khavarian-Garmsir AR, Lassaube U et al (2024) Mapping the implementation practices of the 15-minute city. Smart Cities 7:2094–2109. https://doi.org/10.3390/smartcities7040083

Al-Raeei M (2024) The smart future for sustainable development: Artificial intelligence solutions for sustainable urbanization. Sustain Dev. https://doi.org/10.1002/sd.3131

Anciaes P, Alhassan JAK (2024) Economic and social impacts of public transport investments: a scoping literature review, pp 231–258

Anguelovski I, Shi L, Chu E et al (2016) Equity impacts of urban land use planning for climate adaptation. J Plan Educ Res 36:333–348. https://doi.org/10.1177/0739456X16645166

Argyros D, Jensen AF, Rich J, Dalyot S (2024) Riding smooth: a cost-benefit assessment of surface quality

on Copenhagen's bicycle network. Sustain Cities Soc 108:105473. https://doi.org/10.1016/j.scs.2024.105473
Arias-Meza M, Alvarez-Risco A, Cuya-Velásquez BB et al (2023) Sustainability and urban innovation, pp 81–91
Asgari H, Daudpour Z, Bigdali V, Kermani BM (2024) A comprehensive review of the factors affecting the redefinition of therapeutic use in Metropolitan Cities during the emerging and re-emerging pandemics. Rom J Mil Med 127:357–366. https://doi.org/10.55453/rjmm.2024.127.5.4
Ayesh AN (2024) Enhancing urban living in smart cities using the internet of things (IOT). Int Acad J Sci Eng 11:237–246. https://doi.org/10.9756/IAJSE/V11I1/IAJSE1127
Babic N (2021) Superblocks—the future of walkability in cities? Acad Lett. https://doi.org/10.20935/AL747
Banister D (2008) The sustainable mobility paradigm. Transp Policy 15:73–80. https://doi.org/10.1016/j.tranpol.2007.10.005
Bano S, Hussain A, Arif A et al (2024) Data-driven internet of things: role in smart cities. Asian Bull Big Data Manag 4. https://doi.org/10.62019/abbdm.v4i02.155
Batty M (2013) Big data, smart cities and city planning. Dialogues Hum Geogr 3:274–279. https://doi.org/10.1177/2043820613513390
Bencekri M, Ku D, Lee D et al (2023a) The elasticity and efficiency of carbon reduction strategies in transportation. Energy Sourc Part A Recover Util Environ Eff 45:12791–12807. https://doi.org/10.1080/15567036.2023.2276380
Benton-Short L, Keeley M (2023) Urban sustainability. The Routledge companion to the American Landscape. Routledge, London, pp 267–280
Bocarejo JP, Urrego LF (2022) The impacts of formalization and integration of public transport in social equity: the case of Bogota. Res Transp Bus Manag 42:100560. https://doi.org/10.1016/j.rtbm.2020.100560
Boussaa D, Madandola M (2024) Cultural heritage tourism and urban regeneration: the case of Fez Medina in Morocco. Front Archit Res. https://doi.org/10.1016/j.foar.2024.04.008
Burton E (2000) The Compact city: just or just compact? A preliminary analysis. Urban Stud 37:1969–2006. https://doi.org/10.1080/00420980050162184
Carro-Suárez J, Sarmiento-Paredes S, Nava D (2023) Smart and sustainable cities: a new urban transformation. In: Sustainable regional planning. IntechOpen
Cee K (2023) 15-minute city
Cervero R, Kockelman K (1997) Travel demand and the 3Ds: density, diversity, and design. Transp Res Part D Transp Environ 2:199–219. https://doi.org/10.1016/S1361-9209(97)00009-6
Chen Q (2023) Sustainable community transformation and planning based on community layout and low impact development concept. Highlights Sci Eng Technol 75:156–161. https://doi.org/10.54097/8tb8nb31
Chen S (2024) Planning and development of smart cities based on the concept of sustainable development. Adv Econ Manag Polit Sci 96:73–80. https://doi.org/10.54254/2754-1169/96/2024MUR0115
Congiu T, Sotgiu G, Castiglia P et al (2019) Built environment features and pedestrian accidents: An Italian retrospective study. Sustainability 11:1064. https://doi.org/10.3390/su11041064
Cooper DG (2024) Participatory urban planning. In: The Cambridge handbook of community empowerment. Cambridge University Press, pp 241–261
Dewantara JA (2024) Natural rights on the book "Silent Springer" critical review of ecological citizenship. J Pendidik PKN (Pancasila dan Kewarganegaraan) 5:1. https://doi.org/10.26418/jppkn.v5i1.71751
Díaz CG, Zambrana-Vasquez D, Bartolomé C (2024) Building resilient cities: a comprehensive review of climate change adaptation indicators for urban design. Energies 17:1959. https://doi.org/10.3390/en17081959
Duan L, Sun W, Zheng S (2020) Transportation network and venture capital mobility: an analysis of air travel and high-speed rail in China. J Transp Geogr 88:102852. https://doi.org/10.1016/j.jtrangeo.2020.102852
Dumbaugh E, Li W (2010) Designing for the safety of pedestrians, cyclists, and motorists in urban environments. J Am Plan Assoc 77:69–88. https://doi.org/10.1080/01944363.2011.536101
Echendu AJ (2023) Urban planners' perspectives of public participation in planning in Nigeria. SN Soc Sci 3:33. https://doi.org/10.1007/s43545-022-00604-4
El-Geneidy A, Grimsrud M, Wasfi R et al (2014) New evidence on walking distances to transit stops: identifying redundancies and gaps using variable service areas. Transportation (Amst) 41:193–210. https://doi.org/10.1007/s11116-013-9508-z
Evans G (2014) Living in the city. In: Wellbeing. Wiley, pp 1–28
Ewing R, Cervero R (2001) Travel and the built environment: a synthesis. Transp Res Rec J Transp Res Board 1780:87–114. https://doi.org/10.3141/1780-10
Ewing R, Cervero R (2010) Travel and the built environment. J Am Plan Assoc 76:265–294. https://doi.org/10.1080/01944361003766766
Ezenwa AE, Oledinma A, Osei M et al (2024) Navigating sustainability: a conceptual exploration of road freight decarbonisation. In: Advances in logistics engineering [Working Title]. IntechOpen
Fainstein NI, Fainstein SS (1987) Economic restructuring and the politics of land use planning in New York City. J Am Plan Assoc 53:237–248. https://doi.org/10.1080/01944368708976658
Fejza A (2022) Spatial elements that create the quality public space. Igra ustvarjalnosti - Creat Game 14–21. https://doi.org/10.15292/IU-CG.2022.10.014-021
Fonseca ED (2024) Eco-urbanism: new synergies between architecture, sustainable development, heritage and their multiple visions. ESTOA 13:7–9. https://doi.org/10.18537/est.v013.n025.e01
García García MJG (2022) Renewable energy in the framework of a sustainable urbanism. In: Eco-friendly and

agile energy strategies and policy development, pp 172–189
Garofolo I (2023) The role and implication of UD to foster inclusion in built environments
Geertman S, Stillwell J (2009) Planning support systems: content, issues and trends, pp 1–26
Geetha Varma V (2022) Water-efficient technologies for sustainable development. In: Current directions in water scarcity research, pp 101–128
Geneletti D (2003) Biodiversity impact assessment of roads: an approach based on ecosystem rarity. Environ Impact Assess Rev 23:343–365. https://doi.org/10.1016/S0195-9255(02)00099-9
Geyer HS (2024) The theory and praxis of mixed-use development—an integrative literature review. Cities 147:104774. https://doi.org/10.1016/j.cities.2023.104774
Giannico V, Spano G, Elia M et al (2021) Green spaces, quality of life, and citizen perception in European cities. Environ Res 196:110922. https://doi.org/10.1016/j.envres.2021.110922
Gil-García IC, García-Cascales MS, Dagher H, Molina-García A (2021) Electric vehicle and renewable energy sources: motor fusion in the energy transition from a multi-indicator perspective. Sustainability 13:3430. https://doi.org/10.3390/su13063430
Giudice B, Novarina G, Voghera A (2023) Green infrastructure. Springer International Publishing, Cham
Goerzen A (2024) Global cities. In: Encyclopedia of international strategic management. Edward Elgar Publishing, pp 118–120
Green E (2023) Sustainable development practices in urban planning. https://sigmaearth.com/sustainable-development-practices-in-urban-planning/
Greve M, Laustsen CB (2024) The long durée: a question of sustainability. Politica 56:186. https://doi.org/10.7146/politica.v56i2.147334
Gupta M, Abdelsalam M, Khorsandroo S, Mittal S (2020) Security and privacy in smart farming: challenges and opportunities. IEEE Access 8:34564–34584. https://doi.org/10.1109/ACCESS.2020.2975142
Gürçam S (2023) Paving the way for climate resilience through sustainable urbanization: a comparative study. Lectio Soc. https://doi.org/10.47478/lectio.1396483
Haghani M, Sabri S, De Gruyter C et al (2023) The landscape and evolution of urban planning science. Cities 136:104261. https://doi.org/10.1016/j.cities.2023.104261
Haines A (2000) Fossil fuels, transport, and public health. BMJ 321:1168–1169. https://doi.org/10.1136/bmj.321.7270.1168
Hoel DG (2018) Quantitative risk assessment in the 1970s: a personal remembrance. Dose-Response 16:155932581880323. https://doi.org/10.1177/1559325818803230
Huang Q, Yi C (2024) How does the construction of climate adaptive cities enhance climate resilience? J Environ Plan Manag 1–23 https://doi.org/10.1080/09640568.2024.2362920
Jabareen YR (2006) Sustainable urban forms. J Plan Educ Res 26:38–52. https://doi.org/10.1177/0739456X05285119
Jennings V, Rigolon A, Thompson J et al (2024) The dynamic relationship between social cohesion and urban green space in diverse communities: opportunities and challenges to public health. Int J Environ Res Public Health 21:800. https://doi.org/10.3390/ijerph21060800
Joensuu T, Edelman H, Saari A (2020) Circular economy practices in the built environment. J Clean Prod 276:124215. https://doi.org/10.1016/j.jclepro.2020.124215
Kajiita RM, Kang'ethe SM (2024) Socio-economic dynamics inhibiting inclusive urban economic development: implications for sustainable urban development in South African Cities. Sustainability 16:2803. https://doi.org/10.3390/su16072803
Kapse N, Jangale T (2024) Resilience and sustainability through green infrastructure. J Recent Act Archit Sci 9:6–17. https://doi.org/10.46610/JoRAAS.2024.v09i01.002
Khavarian-Garmsir AR, Sharifi A, Hajian Hossein Abadi M, Moradi Z (2023a) From garden city to 15-minute city: a historical perspective and critical assessment. Land 12:512. https://doi.org/10.3390/land12020512
Khavarian-Garmsir AR, Sharifi A, Sadeghi A (2023b) The 15-minute city: urban planning and design efforts toward creating sustainable neighborhoods. Cities 132:104101. https://doi.org/10.1016/j.cities.2022.104101
Kiranmai N, Aruna Kumari N, Kadari SR (2024) Improved energy efficiency technologies and its applications. Int J Adv Res Sci Commun Technol 32–40. https://doi.org/10.48175/IJARSCT-19104
Knight F (2023) Ebenezer Howard. Oxford University Press, Oxford
Koch M, Grubb M (2023) ■ Agenda 21. The Earthscan reader in sustainable development. Routledge, London, pp 191–193
Kohli A (2024) Community engagement and participation in planning of a city. Int J Multidiscip Res 6. https://doi.org/10.36948/ijfmr.2024.v06i03.23793
Komninos N, Kakderi C, Panori A, Tsarchopoulos P (2019) Smart city planning from an evolutionary perspective. J Urban Technol 26:3–20. https://doi.org/10.1080/10630732.2018.1485368
Lai S, Zoppi C (eds) (2024) Sustainable spatial planning based on ecosystem services. Green Infrastructure and Nature-Based Solutions. MDPI
Lee J (2022) The co-production of environmental regulation and pollution control technology in 1970s America: the controversy over catalytic converters at the U.S. Environmental Protection Agency. Korean J Am Hist 55:115–164. https://doi.org/10.37732/KJAH.2022.55.115
Lee E (2024) Adopting the N-minute city approach to revitalize healthy cities. Acad Urban Heal 3:27–30. https://doi.org/10.59013/auh.2024.3.1.003

Lee S, Bencekri M (2021) Urban form and public transport design. In: Urban form and accessibility. Elsevier, Amsterdam, pp 289–306
Levine RE, Lin C, Xie W (2020) Local financial structure and economic resilience. SSRN Electron J. https://doi.org/10.2139/ssrn.3755560
Li G (2024a) Sustainable development strategies in urban planning. Adv Econ Manag Polit Sci 96:107–111. https://doi.org/10.54254/2754-1169/96/2024MUR0118
Lilli-Drake B (2023) 15 minute cities
Litman T (2022) Evaluating Transportation equity: guidance for incorporating distributional impacts in transport planning. ITE J (Inst Transp Eng) 92:44–49
Long X, Yu H, Sun M et al (2020) Sustainability evaluation based on the three-dimensional ecological footprint and human development index: a case study on the four island regions in China. J Environ Manage 265:110509. https://doi.org/10.1016/j.jenvman.2020.110509
Loo BPY, Tsoi KH, Feng X et al (2024) Cities and urbanization: balancing the environmental and socioeconomic dimensions of sustainability. Adv Sustain Syst 8. https://doi.org/10.1002/adsu.202300401
Mahobia S, Pawar R (2024) Comprehensive review of smart cities using IOT in Coud. Int J Innov Res Comput Commun Eng 12:6309–6319. https://doi.org/10.15680/IJIRCCE.2024.1205194
Maizlish N, Woodcock J, Co S et al (2013) Health cobenefits and transportation-related reductions in greenhouse gas emissions in the San Francisco Bay Area. Am J Public Health 103:703–709. https://doi.org/10.2105/AJPH.2012.300939
Marshall S (2004) Streets and patterns. Routledge
McPhearson T, Andersson E, Elmqvist T, Frantzeskaki N (2015) Resilience of and through urban ecosystem services. Ecosyst Serv 12:152–156. https://doi.org/10.1016/j.ecoser.2014.07.012
Melillo G (2023) 15 minute city in 10 minutes: diving into what they are, quickly
Mirzaee S, Wang Q (2020) Urban mobility and resilience: exploring Boston's urban mobility network through twitter data. Appl Netw Sci 5:75. https://doi.org/10.1007/s41109-020-00316-9
Moradi A, Vagnoni E (2018) A multi-level perspective analysis of urban mobility system dynamics: what are the future transition pathways? Technol Forecast Soc Change 126:231–243. https://doi.org/10.1016/j.techfore.2017.09.002
Moreno C, Allam Z, Chabaud D et al (2021) Introducing the "15-Minute City": sustainability, resilience and place identity in future post-pandemic cities. Smart Cities 4:93–111. https://doi.org/10.3390/smartcities4010006
Moreno C (2024) The 15-minute city: redesigning urban life with proximity to services. Hal Open Sci
Moreno C, Thorne M, Gehl J (2024) The 15-minute city: a solution to saving our time and our planet
Moser G, de Guerra JBSOA, Nunes NA, Francisco THA (2024) Planning and management in cultural preservation and sustainable development in urban contexts. Concilium 24:237–252. https://doi.org/10.53660/CLM-3065-24E28
Mulbery R (2023) The 15-minute city: a Utopian dream or a Dystopian nightmare?
Müller L (2023) Who are we building for? Tracing Universal Design in Urban Development
Mwamba D (2023) The dangers of 15 minute cities: an undemocratic and violation of human rights
NAOCTO (2012) Urban street design guide
Newell R, Picketts I, Dale A (2020) Community systems models and development scenarios for integrated planning: lessons learned from a participatory approach. Commun Dev 51:261–282. https://doi.org/10.1080/15575330.2020.1772334
Nikolov N (2024) Smart cities as a tool for environmental sustainability: opportunities and challenges. Environ Technol Resour Proc Int Sci Pract Conf 1:261–266. https://doi.org/10.17770/etr2024vol1.7948
Oinam S, Chakraborty C (2023) Green IT for sustainable development. In: Smart trends in computing and communications, pp 783–791
Oliveira V (2016) Urban morphology. Springer International Publishing, Cham
Omole FO, Olajiga OK, Olatunde TM (2024) Sustainable urban design: a review of eco-friendly building practices and community impact. Eng Sci Technol J 5:1020–1030. https://doi.org/10.51594/estj.v5i3.955
Panagopoulos T, González Duque JA, Bostenaru Dan M (2016) Urban planning with respect to environmental quality and human well-being. Environ Pollut 208:137–144. https://doi.org/10.1016/j.envpol.2015.07.038
Pandit A, Minné EA, Li F et al (2017) Infrastructure ecology: an evolving paradigm for sustainable urban development. J Clean Prod 163:S19–S27. https://doi.org/10.1016/j.jclepro.2015.09.010
Pavel MEUI (2024) Building equity: exploring the impact of sustainable urban policies on social welfare and inclusivity. Br J Environ Stud 4:18–28. https://doi.org/10.32996/bjes.2024.4.1.3
Perrelet K, Moretti M, Dietzel A et al (2024) Engineering blue-green infrastructure for and with biodiversity in cities. NPJ Urban Sustain 4:27. https://doi.org/10.1038/s42949-024-00163-y
Petit-Boix A, Leipold S (2018) Circular economy in cities: reviewing how environmental research aligns with local practices. J Clean Prod 195:1270–1281. https://doi.org/10.1016/j.jclepro.2018.05.281
Petrie M (2022) Integrating economic and environmental policy. Policy Q 18:10–17. https://doi.org/10.26686/pq.v18i2.7569
Petrov D, Alseghayer R, Chrysanthis PK (2019) Mitigating congestion using environment protective dynamic traffic orchestration. In: 2019 20th IEEE International conference on mobile data management (MDM). IEEE, pp 593–598
Pozoukidou G, Chatziyiannaki Z (2021) 15-minute city: decomposing the new urban planning Eutopia. Sustainability 13:928. https://doi.org/10.3390/su13020928

Pramanik A, Xu P, Xu Y (2023) Equity promotion in public transportation. Proc AAAI Conf Artif Intell 37:11890–11898. https://doi.org/10.1609/aaai.v37i10.26403

Prentice V, Tao J, Bamanie M et al (2024) Municipal sustainability plans and the inclusion of social equity initiatives. J Soc Equity Public Adm 2:1–19. https://doi.org/10.24926/jsepa.v2i2.5509

Prodanovic V, Bach PM, Stojkovic M (2024) Urban nature-based solutions planning for biodiversity outcomes: human, ecological, and artificial intelligence perspectives. Urban Ecosyst. https://doi.org/10.1007/s11252-024-01558-6

Pucher J, Buehler R, Merom D, Bauman A (2011) Walking and cycling in the United States, 2001–2009: evidence from the national household travel surveys. Am J Public Health 101:S310–S317. https://doi.org/10.2105/AJPH.2010.300067

Qian H, Zhang F, Qiu B (2023) Deciphering the evolution, frontier, and knowledge clustering in sustainable city planning: a 60-year interdisciplinary review. Sustainability 15:16854. https://doi.org/10.3390/su152416854

Ramachandra T, Aithal BH, Sanna DD (2012) Insights to urban dynamics through landscape spatial pattern analysis. Int J Appl Earth Obs Geoinf 18:329–343. https://doi.org/10.1016/j.jag.2012.03.005

Ramaiah M, Avtar R (2019) Urban green spaces and their need in cities of rapidly urbanizing india: a review. Urban Sci 3:94. https://doi.org/10.3390/urbansci3030094

Raman R, Roy UK (2019) Taxonomy of urban mixed land use planning. Land Use Policy 88:104102. https://doi.org/10.1016/j.landusepol.2019.104102

Rode P, Floater G, Thomopoulos N et al (2017) Accessibility in cities: transport and urban form, pp 239–273

Rogerson CM (2024) Planning urban regeneration through heritage tourism: the case of Kliptown, South Africa. Mod Geográfia 19:71–89. https://doi.org/10.15170/MG.2024.19.02.05

Rychlewski J (2016) Street network design for a sustainable mobility system. Transp Res Procedia 14:528–537. https://doi.org/10.1016/j.trpro.2016.05.108

Sahu P, Debsarma C (2023) Climate Change and urban environment sustainability: issues and challenges, pp 1–13

Schneider RJ (2013) Theory of routine mode choice decisions: an operational framework to increase sustainable transportation. Transp Policy 25:128–137. https://doi.org/10.1016/j.tranpol.2012.10.007

Sharifi A (2016) From garden city to eco-urbanism: the quest for sustainable neighborhood development. Sustain Cities Soc 20:1–16. https://doi.org/10.1016/j.scs.2015.09.002

Sharma R, Pareek R (2024) Editorial: new paradigms of sustainability in the contemporary era. In: New paradigms of sustainability in the contemporary era. CSMFL Publications, pp iii–iv

Sharma R, Goel S, Lenka SR, Satpathy PR (2024) Energy efficiency retrofitting measures of an institutional building: a case study in eastern India. Clean Energy Syst 7:100111. https://doi.org/10.1016/j.cles.2024.100111

Shaw S-L (2021) Urban human dynamics, pp 41–57

Shoina M, Voukkali I, Anagnostopoulos A et al (2024) The 15-minute city concept: the case study within a neighbourhood of Thessaloniki. Waste Manag Res J Sustain Circ Econ 42:694–710. https://doi.org/10.1177/0734242X241259926

Silva RH, Zwarteveen M, Stead D, Bacchin TK (2024) Bringing ecological urbanism and urban political ecology to transformative visions of water sensitivity in cities. Cities 145:104685. https://doi.org/10.1016/j.cities.2023.104685

Silwall S, Thomas R, Shrestha S (2023) Ecosystem service concept as a rationale for urban green space conservation—a systematic review report. Int J Sci Adv 4. https://doi.org/10.51542/ijscia.v4i3.2

Southon GE, Jorgensen A, Dunnett N et al (2018) Perceived species-richness in urban green spaces: cues, accuracy and well-being impacts. Landsc Urban Plan 172:1–10. https://doi.org/10.1016/j.landurbplan.2017.12.002

Southworth M (2016) Learning to make liveable cities. J Urban des 21:570–573. https://doi.org/10.1080/13574809.2016.1220152

Srikanth AD, Schroepfer T (2023) Network science-based analysis of urban green spaces in Singapore. Int J Smart Sustain Cities 01. https://doi.org/10.1142/S2972426023400044

Stevens MR (2017) Does compact development make people drive less? J Am Plan Assoc 83:7–18. https://doi.org/10.1080/01944363.2016.1240044

Stroope J (2023) Active transportation, context, and community participation: engaged citizens and destination-based walking and biking. Louisiana State University and Agricultural and Mechanical College

Svoboda P (2017) The impact of tax incentives on research and development. Acta Univ Agric Silvic Mendelianae Brun 65:737–743. https://doi.org/10.11118/actaun201765020737

Talen E (2024a) Garden cities. In: What cities say. Oxford University Press, New York, pp 51–56

Taylor AA, Khan MY (2018) Federal air and water regulations. In: Regulatory toxicology, 3rd ed. CRC Press, Boca Raton, pp 265–280

Tedeschi G (2024) Integrating urban energy resilience in strategic urban planning: sustainable energy and climate action plans and urban plans in three case studies in Italy. Land 13:450. https://doi.org/10.3390/land13040450

Tiago P, Leal AI, Silva CM (2024) Assessing ecological gains: a review of how arthropods, bats and birds benefit from green roofs and walls. Environments 11:76. https://doi.org/10.3390/environments11040076

Tu S, Long H, Zhang Y et al (2018) Rural restructuring at village level under rapid urbanization in metropolitan suburbs of China and its implications for innovations in land use policy. Habitat Int 77:143–152. https://doi.org/10.1016/j.habitatint.2017.12.001

Ulak A (2023) Diagrammatic abstractions: Jay Forrester's urban dynamics and its contribution to architecture and urban planning in the late 1960s and early 1970s. Archit Hist 11. https://doi.org/10.16995/ah.8284

Umoh AA, Ohenhen PE, Chidolue O et al (2024) Incorporating energy efficiency in urban planning: a review of policies and best practices. Eng Sci Technol J 5:83–98. https://doi.org/10.51594/estj.v5i1.729

van Egeraat C, Jacobson D (2005) Geography of production linkages in the Irish and Scottish microcomputer industry: the role of logistics. Econ Geogr 81:283–303. https://doi.org/10.1111/j.1944-8287.2005.tb00271.x

Vijay A (2023) Garden cities. In: Topothesia. Fordham University Press, pp 31–59

Wang X, Hawkins C, Berman E (2014) Financing sustainability and stakeholder engagement. Urban Aff Rev 50:806–834. https://doi.org/10.1177/1078087414522388

Webb H (2023) 15 minute city debating its necessity

Whitney RA, Ledsham T (2024) Community animators and participatory planning. J Am Plan Assoc 90:336–348. https://doi.org/10.1080/01944363.2023.2240794

Whittle N (2021) The 15 Minute city: global change through local living

Williams K (2014) Urban form and infrastructure : a morphological review. Futur Cities Work Pap 58

Xiao Z (2024) Beyond a green future: a comprehensive analysis of electrification in transportation—opportunities, challenges, and suggestions. Appl Comput Eng 59:260–267. https://doi.org/10.54254/2755-2721/59/20240817

Yang J (2023) Visualizing and assessing the 15-minute city facility configuration of city region. A study on the Beijing-Tianjin-Hebei Urban Agglomeration. Adv Educ Humanit Soc Sci Res 4:63. https://doi.org/10.56028/aehssr.4.1.63.2023

Yannis G, Chaziris A (2022) Transport system and infrastructure. Transp Res Procedia 60:6–11. https://doi.org/10.1016/j.trpro.2021.12.002

Yeo MT (2019) From garden city to city in a garden and beyond, pp 21–25

Zakari A (2022) The role of green finance in promoting sustainable economic and environmental development. Stud Appl Econ 40. https://doi.org/10.25115/eea.v40i3.5398

Zaman M, Puryear N, Abdelwahed S, Zohrabi N (2024) A review of IoT-based smart city development and management. Smart Cities 7:1462–1501. https://doi.org/10.3390/smartcities7030061

Zhang F, Qian H (2024) A comprehensive review of the environmental benefits of urban green spaces. Environ Res 252:118837. https://doi.org/10.1016/j.envres.2024.118837

Zhou H, Gao H (2020) The impact of urban morphology on urban transportation mode: a case study of Tokyo. Case Stud Transp Policy 8:197–205. https://doi.org/10.1016/j.cstp.2018.07.005

Zulfiqar MU, Kausar M (2023) Historical development of urban planning theory: review and comparison of theories in urban planning. Int J Innov Sci Technol 37–55. https://doi.org/10.33411/IJIST/2023050103

# 15-Minute City Model

# 2

## 2.1 Introduction

The pressing challenges of climate change and urbanization advocate for adopting transformative approaches, such as the 15-minute city framework. This framework is a visionary urban planning model designed to create compact, self-sufficient neighbourhoods where residents can access essential services, such as work, education, health care, and leisure, within a short walk or bike ride (Marchigiani and Bonfantini 2022). By reducing reliance on automobiles, this model not only promotes healthier lifestyles but also fosters equitable economic activity and mitigates environmental harm (Pozoukidou and Angelidou 2022).

Essential components of the 15-minute city include mixed-use zoning, affordable housing, and advanced tools like Geographic Information Systems (GIS) to optimize planning and implementation (Khavarian-Garmsir et al. 2023a). While cities like Paris and Melbourne demonstrate the model's success, challenges such as gentrification and adapting the framework to diverse urban densities remain significant (Gong 2024).

This chapter explores this groundbreaking concept, exploring its theoretical framework, offering insights into its benefits and challenges, and displaying its transformative vision through real-world cases. By synthesizing these elements, this chapter offers a roadmap use the 15-minute city model as paradigm for shaping sustainable and resilient futures cities.

## 2.2 The 15-Minute City Concept

### 2.2.1 Definition

The 15-minute city has emerged as a transformative vision for sustainable urban living, offering a compelling response to the challenges posed by rapid urbanization, environmental degradation, and the need for community-centred planning. At its core, this concept envisions urban spaces designed to ensure that all residents can access essential services (Fig. 2.1), workplaces, schools, health care, markets, and recreational facilities, within a 15-minute walk or bike ride (Moreno et al. 2021). By prioritizing proximity, accessibility, and sustainability, the 15-minute city redefines the way we interact with our built environment, creating neighbourhoods that are compact, interconnected, and inclusive.

This reimagining of city life aims to transcend convenience. This concept seeks to enhance urban resilience and quality of life by fostering social connections and promoting active and healthier lifestyles. Basbas et al. (2023) emphasized its role in nurturing community ties and fostering a health-conscious urban culture. In addition, the



M. Bencekri and S. Lee, *Sustainable Urban Evolution*, Sustainable Development Goals Series,
https://doi.org/10.1007/978-981-96-5146-7_2

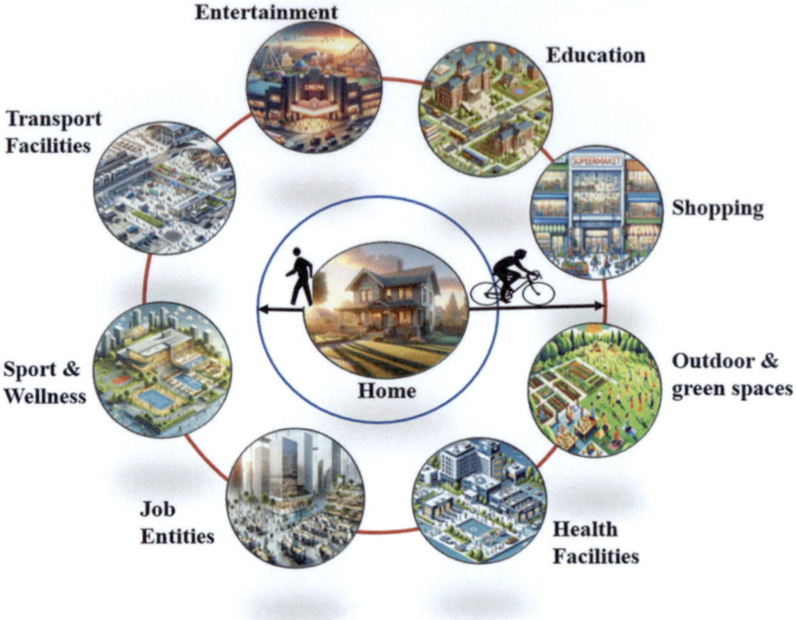


**Fig. 2.1** The 15-minute city concept

15-minute city is inextricably linked to sustainable consumption patterns, presenting a framework for urban development that significantly reduces environmental footprint (Shoina et al. 2024).

### 2.2.2 Key Pillars

Adopting the 15-minute city model requires a fundamental reorientation of urban design priorities, shifting the focus from vehicular traffic to the well-being and needs of residents. This shift not only addresses pressing urban challenges, such as traffic congestion, air pollution, and greenhouse gas emissions, but also significantly enhances the overall quality of life (Jin et al. 2024).

Central to the success of the 15-minute city is active ***community engagement***. Involving residents in urban planning ensures developments align with their needs and preferences, fostering ownership, accountability, and trust. This participatory approach strengthens social cohesion, forming the foundation for inclusive and thriving communities (Shoina et al. 2024; Jin et al. 2024).

Equally important is ***interdisciplinary collaboration***. Effective 15-minute city planning integrates insights from urbanism, sociology, environmental science, and economics to address complex sustainability challenges. For instance, integrating geographical proximity, environmental strategies, land-use economy, and people-centric decision-making, can enhance cities' sustainability, resilience, inclusivity, and liveability (Shoina et al. 2024).

Urban heat islands pose a significant issue, necessitating innovative planning and design practices (Elmarakby and Elkadi 2024). Thus, another critical pillar is the ***environmental and Health considerations***, through the integration of green infrastructure and policies (Allam et al. 2022b). Urban planners are increasingly incorporating elements, such as green roofs and parks, into their designs to enhance biodiversity, reduce heat retention, and provide natural cooling (Xu et al. 2024). These green spaces also serve as recreational areas that promote physical activity and social interaction, which are essential components of a healthy urban lifestyle. For instance, (Manaugh et al. 2015) demonstrated

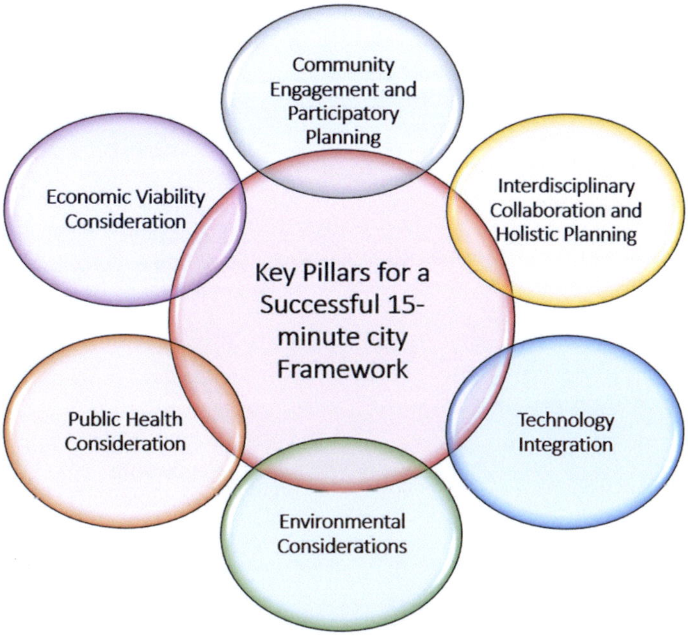


**Fig. 2.2** Key pillars for a successful 15-minute city model

the model's effectiveness in reducing emissions and improving public health outcomes.

***Technological advancements*** play a crucial role in facilitating the transition to a 15-minute city. The use of smart city technologies, such as Geographic Information Systems (GIS) and Internet of Things (IoT) devices, enables more precise and efficient urban planning. These technologies can help monitor and manage urban resources, optimize transportation networks, and provide real-time data on environmental conditions (Kunjir et al. 2023). By leveraging these tools, urban planners can design more responsive and adaptive urban environments that meet residents' evolving needs.

***Economic considerations*** are also integral to the implementation of the 15-minute city. Encouraging local businesses and creating vibrant neighbourhoods' economies can reduce the need for long commutes (Negi et al. 2024) and stimulate local economic growth (Galiautdinov 2023). Policies that support small enterprises, promote mixed-use development, and ensure affordable housing can help create diverse and resilient local economies (Moos et al. 2018). By fostering economic opportunities within walking or cycling distances, the 15-minute city model can enhance the overall economic vitality of urban areas (Lee 2024).

Figure 2.2 outlines the key pillars of a successful implementation of the 15-minute city model, emphasizing six (6) interconnected components: community engagement to align urban planning with residents' needs, interdisciplinary collaboration for integrated designs, and technology integration to optimize infrastructure. It also highlights environmental considerations for sustainability, public health promotion through accessible and walkable spaces, and economic viability to ensure equitable resource distribution and local economic growth. Together, these pillars permits to fulfil the transformative approach of the 15-minute city model, placing accessibility, sustainability, and community well-being at the forefront while moving away from car-dependent infrastructure. Its holistic vision promotes healthier, more resilient, and inclusive neighbourhoods, enhancing the overall quality of life.

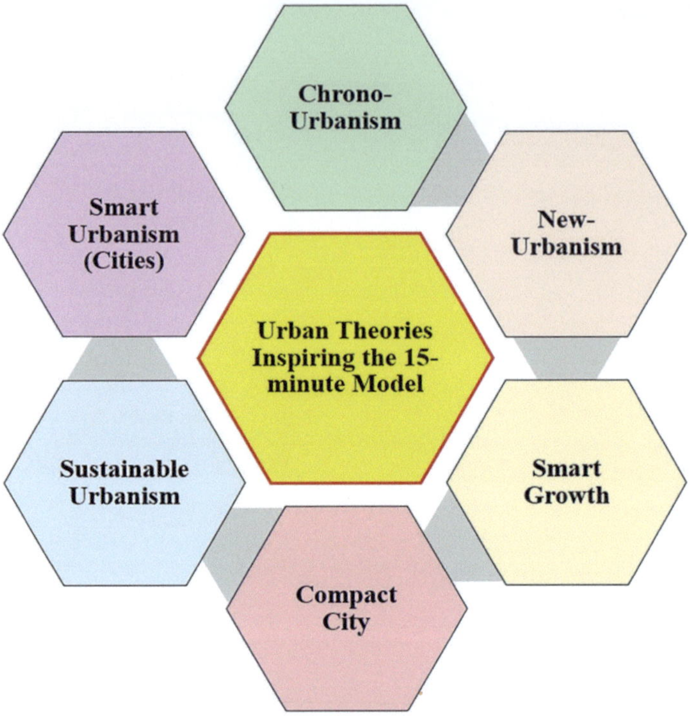


**Fig. 2.3** Urban theories leading to 15-minute city framework

## 2.3 Theoretical Framework

Shoina et al. (2024) emphasized the importance of an interdisciplinary approach in transforming cities, particularly in the context of the 15-minute city concept, considered as a form of sustainable urbanism. The interdisciplinary approach, which is influenced by the smart and green urban transition, helps in crafting effective urban planning policies to address complex challenges such as pollution, poverty, and climate change (Shoina et al. 2024).

Sustainable urbanism advocates for developments that minimize environmental degradation, such as greenhouse gas emissions and energy usage, and promotes a balanced relationship between urban and natural systems (Omole et al. 2024). However, sustainable cities are not only environmentally sustainable, but also conducive to well-being. It emphasizes the provision of sustainable well-being for inhabitants without reducing urban ecosystem services, emphasizing the importance of green infrastructure and low-carbon lifestyles (Li 2024b).

The 15-minute city concept draws inspiration from several urban planning theories (Fig. 2.3).

- **Chrono-urbanism**: Originating from Carlos Moreno's work, chrono-urbanism emphasizes that the significant impact commuting times have on urban quality of life. The 15-minute city addresses this by promoting proximity to amenities and reducing the need for long travel distances, thereby enhancing the quality of urban living (Moreno et al. 2021).
- **New Urbanism**: This theory underscores the importance of walkable neighbourhoods and mixed-use developments (Talen 2024b), which are pivotal to the 15-minute city model. Advocating for compact and mixed development, diverse housing, and pedestrian-oriented designs, it mirrors the 15-minute city's commitment to creating sustainable community-centric living environments (Pozoukidou and Angelidou 2022).
- **Smart Growth**: This approach aims to reduce dependency on automobiles, address

urban sprawl, and environmental degradation through strategic urban planning (Krishnaveni and Anilkumar 2022). Aligning with Smart Growth principles, the 15-minute city focused on sustainable development and resource efficiency.

- **Compact City**: Emphasizing density, mixed-use principles, and efficient public transportation infrastructure, Compact City theory supports the goal of creating sustainable and vibrant urban communities through efficient land use (Bin Sulaiman 2023).
- **Sustainable Urbanism**: A comprehensive approach to address the challenges faced by rapidly growing urban areas, aiming to improve the quality of life of urban residents while minimizing environmental impact (Naku 2024). It aligns with the 15-minute city, of which the primary goals include reducing cars, enhancing energy efficiency, promoting healthier lifestyles through active and public transportation supporting localized economies, and boosting community resilience and liveability (Lee 2024).
- **Smart urbanism**: This phenomenon refers to the integration of advanced technologies and innovative solutions in urban environments to enhance efficiency, sustainability, and quality of life (De Falco 2021). The 15-minute city concept aligns with smart urbanism, as it is part of the broader Smart and Green Transition framework (Shoina et al. 2024).

These theoretical foundations collectively inform the design principles of a 15-minute city, advocating for an urban planning model responsive to the needs of both people and the planet. By synthesizing these diverse yet complementary theories, the 15-minute city model advocates for a city composed of proximate, dense, diverse, digitalized, flexible, connected, and human-scaled neighbourhoods (Fig. 2.4), aligning with the UN SDGs to foster sustainable cities and communities (Khavarian-Garmsir et al. 2023a).

## 2.4 Benefits and Challenges

The 15-minute city directly addresses the shortcomings of transportation systems, which rely heavily on automobiles, leading to traffic congestion, air pollution, and social inequities (Todd 2023). Urban sprawl compounds these issues by making access to employment, education, and healthcare difficult without a car (Barrington-Leigh and Millard-Ball 2015).

By fostering compact, self-sufficient neighbourhoods, the 15-minute city model has emerged as a transformative vision, that departs from traditional car-centric urban planning and offers significant environmental, economic, and social benefits (Allam et al. 2023). Nevertheless, flexible planning frameworks, tailored strategies, and phased implementation approaches are critical for navigating these transitions effectively (Newman et al. 2021).

### 2.4.1 Key Benefits

The 15-minute city framework, as depicted in Fig. 2.5, offers a pathway towards environmental sustainability, economic viability, social equity and inclusivity, and enhanced public health (Allam et al. 2022b), aligning with the Sustainable Development Goals (SDGs), including SDG 11 (Sustainable Cities and Communities), SDG 13 (Climate Action), and SDG 3 (Good Health and Well-Being).

#### 2.4.1.1 Environmental Sustainability

Environmentally, the 15-minute city prioritizes active mobility, which reduces reliance on motorized transport, cuts emissions, and fosters healthier lifestyles (Caselli 2021). Infrastructure such as pedestrian pathways, bike lanes, and safe crossings makes walking and cycling more appealing while encouraging social interactions in public spaces (Kaplan and Giacomo Prato 2015).

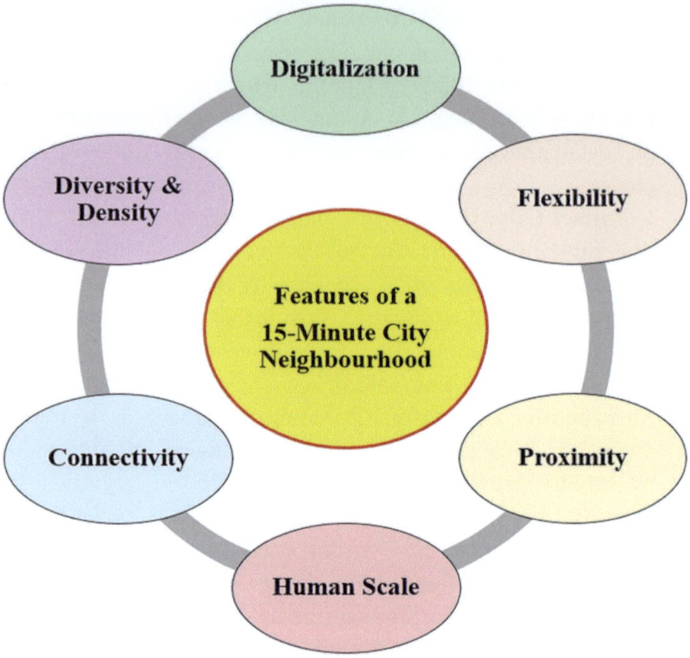


**Fig. 2.4** Design elements of a 15-minute city (Khavarian-Garmsir et al. 2023b)

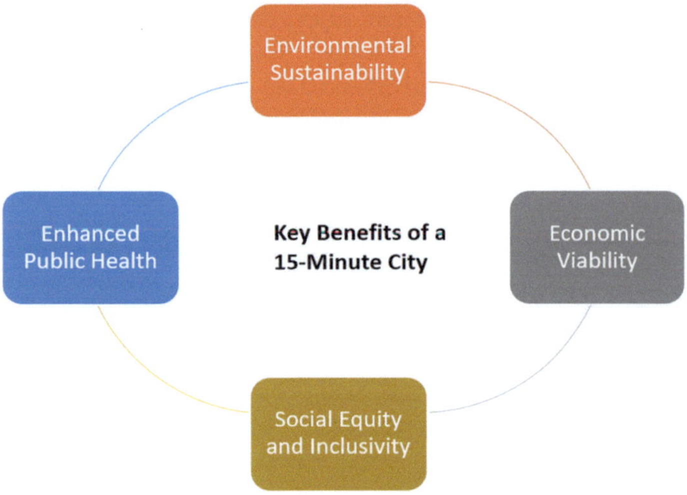


**Fig. 2.5** Key benefits of a 15-minute city model

In cities like Paris, redesigning neighborhoods to expand bike lanes and pedestrian zones significantly reduces traffic congestion and improves environmental sustainability (Moreno et al. 2021). Similarly, whereas, cities like Copenhagen and Amsterdam serve as exemplary models, showing how investments in cycling infrastructure and pedestrian-friendly design can enhance sustainability, alleviate traffic congestion, and improve the environment (Pucher and Dijkstra 2003); (Pucher and Buehler 2008); (Gössling 2013).

Public transportation is another critical environmental element. Reliable, efficient, and affordable public transit systems reduce the need for motorized travel while providing access to essential services and economic opportunities. Investments in high-frequency bus routes, expanded metro systems, and integrated transit hubs not only enhance accessibility but also contribute to environmental sustainability (Ingvardson and Nielsen 2018)). Clean energy solutions, such as electric buses and renewable-powered trains, further advance SDG 13: Climate Action by mitigating greenhouse gas emissions (Rossi and Bianchi 2024).

Shared mobility services and advanced technologies complement these efforts by providing flexible and low-emission transportation options (Martin 2024). Car-sharing, bike-sharing, and ride-hailing apps reduce road congestion and promote sustainable mobility (Krishnan et al. 2024). Smart traffic management, real-time public transit tracking, and data-driven analytics optimize transportation networks, ensuring efficient and reliable mobility.

By integrating active mobility, sustainable public transit, and advanced technologies, the 15-minute city aligns with SDG 11: Sustainable Cities and Communities, and SDG 13: Climate Action. This holistic approach not only addresses pressing environmental challenges but also supports the creation of liveable, resilient, and environmentally friendly urban environments.

#### 2.4.1.2 Economic Viability

Economically, the 15-minute city model strengthens local economies by creating job opportunities and supporting small businesses, directly contributing to SDG 8: Decent Work and Economic Growth (Yang 2023). The model's mixed-use approach distributes economic opportunities throughout the city, reducing congestion and promoting balanced urban growth (Geyer 2024). Additionally, the model revitalizes neighbourhood economies by increasing pedestrian and cycling travel, ensuring vibrant and self-sufficient communities (Shoina et al. 2024).

Affordable housing policies and inclusionary zoning also play a critical role in sustaining economic vitality (Akinsulire et al. 2024). These policies prevent displacement and ensure that diverse socioeconomic groups can access the benefits of compact, mixed-use neighbourhoods. This approach not only stimulates local economic activity but also strengthens the resilience of urban economies in the face of external shocks. For instance, Melbourne's 20-minute neighbourhoods initiative integrates affordable housing with local amenities, creating balanced and inclusive communities (Pozoukidou and Angelidou 2022).

#### 2.4.1.3 Social Equity and Inclusivity

Socially, the 15-minute city fosters inclusivity by ensuring equitable access to essential services, aligning with SDG 10: Reduced Inequalities (Akrami et al. 2024). By integrating green policies, such as urban parks and eco-friendly buildings, it enhances biodiversity and resilience while contributing to SDG 15: Life on Land and SDG 11: Sustainable Cities and Communities (Filazzola et al. 2019).

Decentralizing urban services also helps distribute economic activity more evenly across the city, easing pressure on central business districts and boosting local economies (Ibrahim 2024). Accessible public spaces and affordable housing initiatives further promote inclusivity while preventing gentrification and displacement (Allam et al. 2022a). This increased accessibility can lead to improved health outcomes, better educational attainment, and enhanced overall well-being for underserved populations (Watkins and Mindell 2010; Neutens 2015; van Schalkwyk and Mindell 2018).

Participatory planning is also a critical component that plays a vital role in advancing social inclusion by involving community members and different stakeholders, in shaping their urban environments. Empowering residents to contribute to decision-making fosters a sense of ownership and ensures that urban development aligns with local needs and values (Kohli 2024).

Furthermore, the 15-minute city model enhances community cohesion and social capital by encouraging local interactions and fostering a sense of belonging among the residents. The creation of walkable neighbourhoods with accessible public spaces, parks, and community centres facilitates social interaction and community engagement (Leyden 2003). These infrastructures support the development of strong social networks that are essential for resilient communities. Empirical studies have shown that neighbourhoods designed with these principles in mind experience lower crime rates and higher levels of civic participation (Alonso et al. 2019).

#### 2.4.1.4 Enhanced Public Health

A critical aspect of the 15-minute city is its alignment with Sustainable Development Goal 3 (Good Health and Well-Being). Cities adopting this approach have reported significant improvements in public health, including reduced obesity rates and enhanced mental well-being. These outcomes are largely attributed to increased physical activity and decreased air pollution, emphasizing the strong relationship between the design principles of the 15-minute city and healthier urban living environments (Lu and Diab 2023).

The 15-minute city concept encourages active transportation, such as walking and cycling, by designing urban spaces where residents can access essential services within a short distance. This shift away from car dependency reduces obesity rates and lowers the prevalence of chronic illnesses, such as diabetes and cardiovascular diseases (Shilton et al. 2024). Furthermore, diminished air pollution from decreased vehicle emissions enhances respiratory health and overall public well-being (Sallis et al. 2016).

Mental health also can benefit significantly from the 15-minute city model. Access to green spaces, reduced commuting times, and increased opportunities for community engagement lower stress levels and improve mental health outcomes (Coldwell and Evans 2018). The availability of local amenities and strong social infrastructure fosters community bonds, reduces isolation, and promotes social cohesion (Alonso et al. 2019).

The 15-minute city presents not only a visionary framework for sustainable urban development but also a practical pathway for enhancing public health. However, its implementation requires addressing inherent challenges and limitations to ensure its equitable success.

### 2.4.2 Key Challenges

The 15-minute city model offers a visionary framework for creating sustainable, equitable, and accessible urban environments. However, transitioning from traditional urban planning to this innovative paradigm presents complex challenges (Martin 2024), as depicted in Fig. 2.6.

These challenges encompass gentrification risks, disparities in urban density, infrastructure inadequacies, cultural and geographical barriers, and environmental trade-offs. Successfully navigating these obstacles requires a nuanced and collaborative approach.

#### 2.4.2.1 Gentrification

A significant concern associated with the 15-minute city is the potential for gentrification (Casarin et al. 2023). Concerns have been raised that affluent areas could attract a disproportionate share of investment and development, resulting in the patchwork of well-serviced enclaves alongside areas with stark disparities in service quality and accessibility (Aristizábal et al. 2023). Such uneven development patterns could deepen urban inequality, contradicting the model's objective of fostering cohesive and equitable urban environments (Pozoukidou and Chatziyiannaki 2021).

Moreover, as neighbourhoods become more accessible and desirable, rising property values and living costs may displace lower-income residents, undermining the model's inclusive vision (Pennington 2021). Without safeguards, these transformations could deepen urban inequality, turning revitalized areas into enclaves of exclusivity (Casarin et al. 2023). This unintended consequence threatens to marginalise communities that the 15-minute city model aims to support and uplift, jeopardising its long-term viability.

#### 2.4.2.2 Infrastructure Retrofitting

The feasibility of implementing the 15-minute city varies significantly based on urban density. High-density cities with robust public transit systems and existing infrastructure are more readily adaptable (Gong 2024). In contrast, sprawling, low-density areas face logistical and financial hurdles in retrofitting car-centric designs into walkable, service-rich environments. This disparity underscores the need for customized solutions that align

**Fig. 2.6** Key challenges of the 15-minute city model

with the structural realities of different urban settings.

The adequacy of infrastructure to support a 15-minute city is another crucial challenge. Retrofitting existing urban landscapes to embody the 15-minute city concept is a formidable challenge that requires a comprehensive shift in urban planning and design philosophy (Shoina et al. 2024).

Another issue to this transition is the modification of zoning laws to facilitate mixed-use development and the reconfiguration of transportation infrastructure to prioritize non-motorized and public transit options over private vehicles (Jin et al. 2024). Additionally, urban design must evolve to support pedestrian-friendly environments and promote walking and cycling.

This urban and infrastructure transformation necessitates substantial financial resources, solid political commitment, and community engagement, as it often entails navigating complex bureaucratic and regulatory landscapes (Allam et al. 2022a).

#### 2.4.2.3 Cultural and Geographical Barriers

The universal application of the 15-minute city model is challenged by the diversity of urban contexts. Cities with unique cultural, historical, or geographical characteristics may find it difficult to adapt a one-size-fits-all approach (Jin et al. 2024). For example, historical city layouts or extreme weather conditions may limit the practicality of cycling or walking as primary modes of transportation (Marchigiani and Bonfantini 2022). Overlooking these nuances and contextual differences risks alienating communities and diminishing the model's effectiveness and relevance.

#### 2.4.2.4 Political and Community Resistance

Realizing the 15-minute city model may encounter resistance from political entities, property developers, businesses, and even residents, particularly in areas where car reliance is prevalent (Shoina et al. 2024). Overcoming such barriers requires strategic stakeholder engagement, emphasizing the importance of community involvement and transparency throughout planning and decision-making processes. Engaging stakeholders early and continuously can help cultivate a shared vision for a sustainable urban future, fostering broader acceptance of and support for transformative projects (Djalali et al. 2020).

#### 2.4.2.5 Resources Allocation

Achieving a 15-minute city model relies on the strategic deployment of resources, including financial, human, and technological assets (Pozoukidou and Chatziyiannaki 2021). For instance, retrofitting demands significant investment in public transit, pedestrian-friendly spaces, and mixed-use developments ((Pardo-Bosch et al. 2019). Especially in cities and suburbs designed around car dependency, achieving a 15-minute city necessitates extensive, potentially costly urban redesigns (Martin 2024).

Funding is critical for infrastructure improvements and the enhancement of public transportation systems, necessitating innovative financing strategies such as public–private partnerships (Delmon 2024). These issue prompt questions about the economic viability of the 15-minute city model, especially where urban sprawl is a foundational aspect of a city's structure (Allam et al. 2022a).

Additionally, skilled labour for planning, execution, and technology for data-driven urban management is crucial for the success of the model. Effective resource allocation can facilitate the implementation of smart city technologies, offering long-term efficiency and sustainability benefits (Allam et al. 2022a).

#### 2.4.2.6 Environmental Trade-Offs

While a 15-minute city aims to reduce carbon emissions and enhance urban green spaces, the construction activities required for its development may temporarily increase pollution and disrupt ecosystems. Employing green construction practices and adhering to sustainable development principles are essential for mitigating these short-term environmental impacts (Pavate et al. 2024). Sustainable planning, alongside rigorous environmental impact assessments, ensures that the model's ecological goals are achieved without compromising its foundational sustainability ethos (Marchigiani and Bonfantini 2022).

### 2.4.3 Mitigation Policies

To address the challenges associated with the 15-minute city concept effectively, urban planners and policymakers must embrace strategies that are both adaptable and attuned to the specific needs and characteristics of each urban environment. The following strategies (Fig. 2.7) provide a roadmap for navigating these complexities and challenges.

#### 2.4.3.1 Anti-gentrification Measures

Mitigating the risk of gentrification is essential to ensure that the 15-minute city benefits all residents, not just affluent groups. Cities should prioritize robust affordable housing policies, such as rent controls, subsidies for low-income residents, and incentives for mixed-income housing development.

Inclusionary zoning and community land trusts can help stabilize neighbourhoods and maintain socioeconomic diversity (Weller and Hulten 2012). By ensuring affordable housing options, cities can prevent displacement and distribute the benefits of urban improvements equitably.

#### 2.4.3.2 Comprehensive Infrastructure Investments

Significant investments in public transportation, cycling paths, pedestrian walkways, and green spaces are critical to creating the infrastructure needed for a functional 15-minute city (Magrinyà et al. 2023). These developments should prioritize underserved neighbourhoods to reduce inequalities and promote cohesive urban development.

Additionally, integrating smart city technologies can enhance infrastructure efficiency, sustainability, and adaptability (Allam et al. 2022c). Ensuring that infrastructure keeps pace with urban growth is vital for maintaining the liveability and accessibility of these reimagined spaces.

**Fig. 2.7** Example of mitigation policies

#### 2.4.3.3 Localized Pilot Projects

Launching localized pilot projects allows cities to test and refine the 15-minute city model in real-world contexts. These small-scale initiatives enable planners to identify and address localized challenges, gather resident feedback, and demonstrate tangible benefits, such as reduced traffic congestion and improved public health. Community engagement during these projects fosters trust and ensures that solutions align with residents' needs (Pozoukidou and Angelidou 2022).

#### 2.4.3.4 Context-Sensitive Urban Planning

Recognizing the cultural, geographical, and historical diversity of cities is critical to successfully implementing the 15-minute city model. Urban planning efforts should account for local norms, environmental conditions, and community priorities, ensuring that the model enhances the existing urban fabric without disrupting it (Marchigiani and Bonfantini 2022). Tailored approaches that reflect local identities increase the model's acceptance and ensure its relevance across diverse contexts.

#### 2.4.3.5 Community Engagement

Proactive and transparent community involvement is crucial for overcoming resistance and achieving equitable outcomes. Stakeholders should be engaged throughout the planning and implementation processes through public consultations, participatory budgeting, and accessible communication. Such engagement fosters trust and aligns policy objectives with the aspirations of the community, promoting a shared vision for urban transformation (Djalali et al. 2020).

#### 2.4.3.6 Sustainable Construction Practices

To mitigate the environmental impacts of urban transformations, cities should prioritize sustainable construction practices. This includes using recycled and energy-efficient materials, conducting rigorous environmental impact assessments, and promoting ecological restoration where necessary. These practices minimize short-term disruptions while aligning urban redevelopment with long-term sustainability goals (Pavate et al. 2024).

### 2.4.4 Recapitulation

The concept of the ***15-minute city*** aims to transform urban environments by addressing the shortcomings of car-dependent urban planning, which leads to traffic congestion, air pollution, and social inequities. By creating compact, self-sufficient neighbourhoods, this model promotes environmental, economic, social, and health benefits while aligning with the diverse SDGs goals. However, its implementation presents notable challenges that require tailored, inclusive, and sustainable approaches.

The ***key benefits*** of the model include environmental sustainability, economic viability, social equity and inclusivity, and enhanced public health. However, several ***key challenges*** are identified, such as gentrification risks, urban and infrastructure retrofitting, cultural and geographical barriers, political and community resistance, resource allocation, and environmental trade-offs. And to address these challenges, the following ***mitigation measures*** are proposed:

- Implement anti-gentrification policies (e.g. rent controls, inclusionary zoning).
- Invest in infrastructure for active mobility and public transit, prioritizing underserved areas.
- Test localized pilot projects to refine the model and gather community feedback.
- Adapt plans to reflect the cultural and geographical diversity of urban contexts.
- Foster community engagement through participatory planning and transparent communication.
- Emphasize sustainable construction practices to align redevelopment with environmental goals.

The 15-minute city represents a visionary yet practical framework for sustainable urban transformation. Success depends on equitable, inclusive, and locally adapted strategies that balance immediate challenges with long-term sustainability objectives. Therefore, examining global case studies is essential to understand how cities worldwide are adapting the 15-minute city model to their unique needs and challenges.

## 2.5 Successful Use Cases

The 15-minute city concept heralds a transformative blueprint for urban development, aiming to revolutionize how cities are organized by ensuring that residents have access to essential services and amenities within a short walk or cycle from their homes. This innovative approach to sustainable mobility not only promises an enhanced quality of life but also contributes to environmental preservation and fosters cohesive communities.

The success of following cities, such as Paris, in implementing the 15-minute city model highlights the scalability and adaptability of this model to other urban contexts (Barbieri et al. 2023).

### 2.5.1 Paris, France

Paris has undergone a significant urban transformation under Mayor Anne Hidalgo (Fig. 2.8), with initiatives such as converting schoolyards into communal parks and expanding the cycling infrastructure (Thaury et al. 2024). These efforts have revitalized local economies, reduced car dependence, and improved environmental quality. However, a 2020 City of Paris report highlighted that these benefits have not been equitably distributed, underscoring the need for more inclusive urban planning (Crook 2021).

Transforming schoolyards into communal parks has provided essential green spaces that enhance neighbourhood aesthetics and improve residents' well-being by increasing access to nature (Rigolon et al. 2015). These parks also serve as community hubs, fostering social cohesion, and offering safe and accessible spaces for families.

Expanding the cycling infrastructure has significantly reduced car usage, reduced CO2 emissions, and improved air quality (Brand et al. 2021). The development of extensive bike lanes and bike-sharing programs, such as Vélib (Fig. 2.9), has made cycling a convenient transportation option, aligning with global sustainability goals (Álvarez et al. 2023). However, the study noted that lower-income neighbourhoods

**Fig. 2.8** Paris 15-minute city concept proposed by Anne Hidalgo (Crook 2021)

did not receive the same level of investment in cycling infrastructure as wealthier areas did (Nair et al. 2013). This disparity highlights the importance of ensuring that sustainable urban initiatives benefit residents equally (Ivanovic et al. 2023). To address these inequities, Paris must prioritize targeted investments in underserved areas, enhance community engagement in planning, and ensure affordable access to sustainable transportation options such as bike-sharing. By focusing on inclusivity, Paris could create a more balanced and just urban environment.

The economic benefits of these initiatives emphasize the link between urban planning and local economies. By creating vibrant public spaces and improving transport infrastructure, Paris attracted businesses, boosted commerce, and supported job creation, enhancing residents' quality of life (Porter et al. 2020).

In conclusion, Paris' urban revitalisation under Mayor Hidalgo demonstrates the environmental and economic benefits of sustainable urban planning. However, to ensure that these benefits are shared equally, the city must adopt a more inclusive approach. By addressing these challenges, Paris can become a model for other cities aiming for equitable and sustainable urban transformation.

### 2.5.2 Melbourne, Australia

The Victorian Government's 20-minute neighbourhood strategy in Melbourne is a forward-thinking urban planning initiative designed to cultivate sustainable and resilient communities. This strategy emphasizes the importance of mixed-use zoning and enhanced public transportation networks with the aim of reducing dependence on private vehicles and promoting local living (Morley and Pafka 2023). A key principle underpinning this approach is the concept of "density done well," which seeks to increase urban density in a manner that is both human-centred and environmentally sustainable, thereby enhancing the overall urban liveability.

The implementation of this strategy has yielded several positive outcomes. By ensuring that essential services, employment opportunities, and recreational spaces are accessible within a 20-minute radius, the strategy has strengthened local

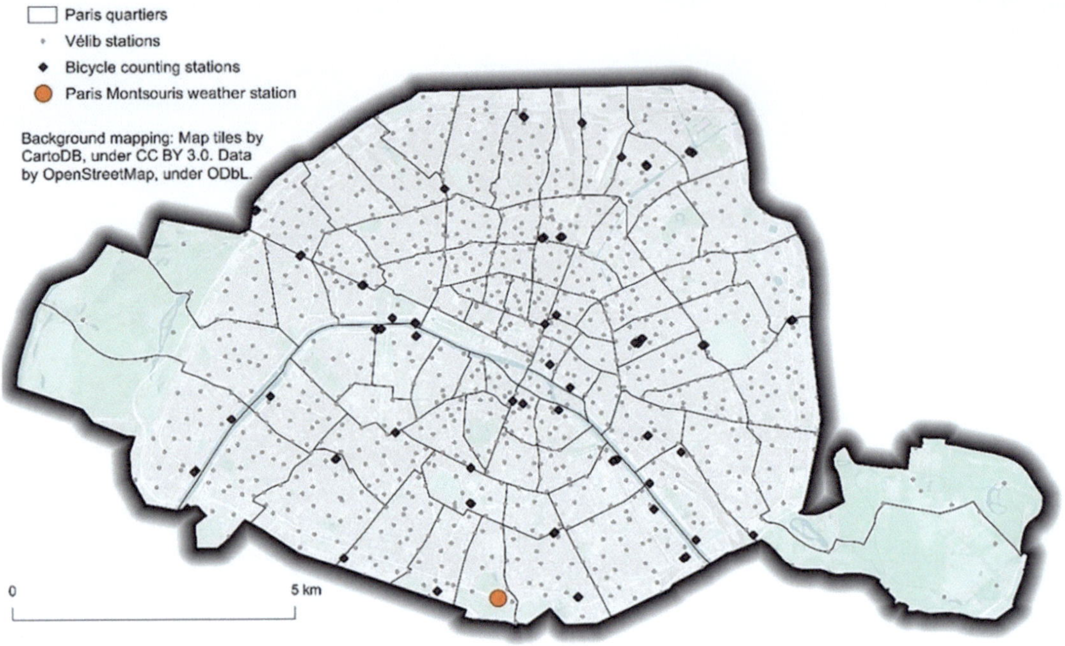


**Fig. 2.9** Vélib Stations in Paris (Ivanovic et al. 2023)

economies and fostered a greater sense of community among the residents. This approach has also led to heightened community satisfaction, as people enjoy the convenience and social cohesion that come with living in well-designed, accessible neighbourhoods.

However, significant challenges remain, particularly in retrofitting the established suburbs to align with this new model. Many of these areas have existing infrastructure and urban designs that are not easily adaptable to high-density living. To overcome these obstacles, the strategy emphasizes the need for flexible zoning regulations that can be tailored to the unique needs of different neighbourhoods. Moreover, the provision of incentives for developers is crucial to encouraging investment in these transformations, making the shift towards a more sustainable urban form both feasible and attractive.

A study conducted by Thornton et al. (2022a; b) operationalized the 20MN concept for Melbourne and Adelaide using data across five domains: healthy food, recreational resources, community resources, public open space, and public transport. Areas where all five domains intersect are considered 20MNs, with more such areas found near city centres and around train stations in Melbourne.

Thornton et al. (2022a; b) provided a table outlining the key components of Melbourne's 20-minute neighbourhood (20MN) concept (Table 2.1). It details six domains: healthy food, recreational resources, community resources, public open spaces, and public transport, each with specific access measures. These elements are crucial for fostering sustainable, liveable urban environments that reduce reliance on cars and enhance residents' quality of life by ensuring that daily needs are within proximity.

By balancing the demands of growth with the preservation of Melbourne's established communities' unique character, the Victorian Government aimed to create a more sustainable and liveable urban future. This strategy not only addresses current urban challenges but also sets a precedent for future-oriented urban development that prioritizes both people and the environment.

**Table 2.1** Melbourne's 20-minute city features (Thornton et al. 2022b)

| Domain | Attributes | Rationale for attribute selection | Access measure |
|---|---|---|---|
| Healthy food | • Large supermarket<br>• Smaller supermarket and greengrocer | Ensures access to affordable, healthy food options within a walkable distance | 1.5 km pedestrian network distance to a large supermarket<br>OR<br>1.5 km to a smaller supermarket and greengrocer |
| Recreational resources | • Gym | Provides access to fitness facilities for various age groups within the community | 1.5 km pedestrian network distance to a gym |
| Community resources | • Primary school<br>• GP<br>• Pharmacy<br>• Library<br>• Post office<br>• Café | Essential services that support education, health, social interaction, and daily needs | 1.5 km pedestrian network distance to all listed community resources |
| Public open space | • Natural and semi-natural spaces<br>• Parks and gardens<br>• Sports fields | Promotes access to public spaces for recreation and relaxation within the community | 400 m pedestrian network distance to public open space<br>AND<br>8 ha of public open space within 1 km |
| Public transport | • Bus stop<br>• Tram stop<br>• Train station | Ensures connectivity and reduces car dependency through accessible public transport | 400 m to bus stop<br>OR<br>600 m to tram stop<br>OR<br>800 m to train station |

### 2.5.3 Portland, Oregon, USA

Portland's urban planning strategies closely align with the 15-minute city concept, particularly through its concept of a 20-minute neighbourhood, which is designed as a complete neighbourhood where residents of all ages can safely and conveniently access everyday amenities by walking or cycling. The city encourages residential development and mixed land use towards neighbourhood centres to make walking, cycling, and public transit more feasible than driving (Rao et al. 2024).

Portland's urban design has effectively incorporated these principles, leading to significant reductions in car dependency and the associated carbon emissions. The neighbourhoods are connected to the central city and other job centres via public transit and protected bike lanes, promoting a more walkable urban environment. The city recognizes 400–800 m as the range for 20-minute walking distances, with a community public space surrounded by various amenities (Rao et al. 2024).

However, Portland suffers from critical concerns about the unintended social consequences of its urban policies and transformations, particularly regarding gentrification and displacement of long-standing communities (Jordan 2013). Gentrification involves the transformation of a working-class area into a middle- or upper-class neighbourhood, often driven by urban renewal initiatives that raise property values and living costs, leading to the displacement of long-term, low-income residents (Mehdipanah et al. 2018). This process can have negative health impacts, such as increased stress and mental health issues, owing to the disruption of social networks and reduced access to resources. Additionally, gentrification exacerbates social exclusion, particularly for low-income, racial, and ethnic minorities, and seniors, who may struggle to relocate and often end up in areas with fewer opportunities (Mehdipanah et al. 2018).

In conclusion, Portland's 20-minute neighbourhoods aim to enhance walkability, cycling, and public transit use by promoting residential development and mixed land use near neighbourhood centres. However, implementing these spatial arrangements across different scales poses challenges, particularly in ensuring connectivity to the central city and other activity centres, which demands significant infrastructure planning and investment (Rao et al. 2024). Balancing the intensity of development while preserving community character is also complex, as efforts to make walking and cycling more feasible can impact existing neighbourhood identities. Additionally, the creation of desirable walkable areas carries the risk of gentrification, potentially displacing long-standing communities due to rising property values and living costs.

### 2.5.4 Recapitulation

The 15-minute city offers a transformative vision for urban development by rethinking traditional car-centric planning and fostering compact, self-sufficient neighbourhoods. This model integrates environmental sustainability, economic vitality, and social equity, providing accessible services within walking or cycling distance and aligning with SDG 11, SDG 13, and SDG 8.

Environmentally, the model reduces emissions and improves urban health through active mobility, reliable public transit, and clean energy solutions. Economically, it revitalizes local economies by supporting small businesses and promoting mixed-use developments that create vibrant, resilient communities. Socially, inclusive policies, participatory planning, and equitable access to green spaces strengthen social cohesion and prevent displacement.

Real-world examples from cities like Paris and Melbourne proved the adaptability and potential of the 15-minute city to improve urban liveability. Paris, for instance, has expanded bike lanes and pedestrian zones to reduce traffic and improve air quality, while Melbourne's 20-minute neighbourhoods have enhanced social connectivity and local economies through mixed-use zoning. Despite these successes, challenges such as gentrification, infrastructure limitations, and varying local contexts emphasize the need for flexible, tailored, and community-centred strategies.

## 2.6 Potential Unlocking

The success of the 15-minute city model is heavily context-dependent, shaped by factors such as sociocultural priorities, economic conditions, and political leadership. Infrastructure funding and neighbourhood economic diversity also significantly influence outcomes, as does political will, exemplified by Paris's decisive leadership in reshaping its urban fabric.

This analysis underscores that while the 15-minute city holds immense potential for fostering sustainable, liveable, and equitable urban environments, its implementation is complex. To fully unlock the transformative power of this model, a holistic and strategic approach is essential. The following recommendations (Fig. 2.10) offer actionable insights for policymakers, urban planners, and communities.

- Ensure Equitable Access
  - Affordable Housing: Implement robust mixed-income housing policies and targeted subsidies to ensure affordability near essential services.
  - Equitable Mobility: Expand subsidies for public transit and active mobility options for low-income residents, ensuring that economic barriers do not hinder accessibility.
- Invest in Active and Public Transportation
  - Infrastructure Priorities: Allocate funding for dedicated bike lanes, pedestrian-friendly zones, and eco-friendly public

**Fig. 2.10** Actionable insights for a 15-minute city framework

transit systems, as exemplified by Copenhagen's cycling infrastructure.
  - Incentives for Transition: Introduce tax breaks or subsidies for businesses and residents that adopt sustainable practices, such as reducing reliance on private vehicles.
- Strengthen Local Economies
  - Small Business Support: Provide grants and low-interest loans for start-ups in mixed-use neighbourhoods. Encourage local procurement policies to stimulate neighbourhood economies.
  - Vibrant Public Spaces: Design community hubs that integrate retail, services, and recreational areas to foster local economic activity.
- Prioritize Green Urban Design
  - Green Spaces: Mandate public green spaces in all new developments. Initiatives like urban tree planting, rooftop gardens, and biodiversity corridors enhance environmental and social benefits.
  - Community Gardens: Establish community gardens to improve food security and promote social interaction.
- Foster Community Engagement
  - Participatory Planning: Form local urban planning committees that include diverse community stakeholders. Use digital platforms to broaden engagement and gather feedback.
  - Empowerment Through Inclusion: Ensure residents have a voice in urban projects, fostering a sense of ownership and trust in the development process.
- Leverage Smart Technologies
  - Data-Driven Solutions: Use technologies like real-time traffic management, energy monitoring, and service optimization to enhance urban efficiency.
  - Public–Private Partnerships: Collaborate with private firms to pilot innovative smart city projects, ensuring that technological advancement aligns with community needs.
- Adopt Flexible Zoning Laws
  - Support Mixed-Use Development: Revise zoning regulations to allow dynamic allocation of urban spaces, promoting mixed-use neighbourhoods.

  - Adaptive Reuse: Encourage the transformation of underutilized buildings into community-centric spaces.
- Promote Green Building Practices
  - Sustainability Standards: Mandate energy-efficient designs for new constructions and incentivize retrofitting of older buildings with green technologies.
  - Low-Impact Construction: Encourage the use of recycled materials and eco-friendly construction practices.
- Innovate Financing Mechanisms
  - Diversify Funding: Explore green bonds, community funding, and public–private partnerships to finance urban redevelopment projects.
  - Tax Incentives: Offer tax breaks for sustainable development projects and businesses prioritizing local hiring and environmental practices.
- Enhance Public Awareness and Education
  - Awareness Campaigns: Launch campaigns to educate residents on the benefits of the 15-minute city, focusing on sustainability and well-being.
  - Urban Education: Incorporate sustainability principles into school curricula to nurture future generations of environmentally conscious citizens.
- Establish Continuous Monitoring
  - Progress Tracking: Create dedicated observatories to monitor project outcomes, share best practices, and adjust strategies as needed.
  - Global Collaboration: Foster knowledge exchange among cities to accelerate the adoption of successful models.

The 15-minute city offers a transformative pathway to create vibrant, inclusive, and sustainable urban spaces. By adopting these recommendations, cities can address current challenges while ensuring resilience against future crises. Through collaboration, innovation, and community engagement, urban environments can evolve into thriving ecosystems that prioritize equity, sustainability, and well-being for all residents.

## References


Akinsulire AA, Idemudia C, Okwandu AC, Iwuanyanwu O (2024) Economic and social impact of affordable housing policies: a comparative review. Int J Appl Res Soc Sci 6:1433–1448. https://doi.org/10.51594/ijarss.v6i7.1333

Akrami M, Sliwa MW, Rynning MK (2024) Walk further and access more! Exploring the 15-minute city concept in Oslo, Norway. J Urban Mobil 5:100077. https://doi.org/10.1016/j.urbmob.2024.100077

Allam Z, Bibri SE, Chabaud D, Moreno C (2022a) The theoretical, practical, and technological foundations of the 15-minute city model: proximity and its environmental, social and economic benefits for sustainability. Energies 15:6042. https://doi.org/10.3390/en15166042

Allam Z, Bibri SE, Chabaud D, Moreno C (2022b) The '15-Minute City' concept can shape a net-zero urban future. Humanit Soc Sci Commun 9:126. https://doi.org/10.1057/s41599-022-01145-0

Allam Z, Bibri SE, Jones DS et al (2022c) Unpacking the '15-Minute City' via 6G, IoT, and digital twins: towards a new narrative for increasing urban efficiency, resilience, and sustainability. Sensors 22:1369. https://doi.org/10.3390/s22041369

Allam Z, Chabaud D, Gall C et al (2023) Redefining investable infrastructure in developing nations in a postpandemic era: the case of the 15-Minute City. In: Allam Z, Chabaud D, Gall C et al (eds) Resilient and sustainable cities. Elsevier, Amsterdam, pp 135–146

Alonso JM, Andrews R, Jorda V (2019) Do neighbourhood renewal programs reduce crime rates? Evidence from England. J Urban Econ 110:51–69. https://doi.org/10.1016/j.jue.2019.01.004

Álvarez LL, Montoya LHB, D'Alleman JC, Nicola Stepanie Klocke NI (2023) Bike sharing systems and the collaborative economy: a systematic literature review, pp 279–297

Aristizábal JE, Sarache W, Escobar DA (2023) Spatial regression model of urban walkability under the 15-minute city approach. Geoj Tour Geosites 49:1037–1045. https://doi.org/10.30892/gtg.49319-1103

Barbieri L, D'Autilia R, Marrone P, Montella I (2023) Graph representation of the 15-minute city: a comparison between Rome, London, and Paris. Sustainability 15:3772. https://doi.org/10.3390/su15043772

Barrington-Leigh C, Millard-Ball A (2015) A century of sprawl in the United States. Proc Natl Acad Sci 112:8244–8249. https://doi.org/10.1073/pnas.1504033112

Basbas S, Campisi T, Papas T, et al (2023) The 15-minute city model: the case of sicily during and after COVID-19. Commun Sci Lett Univ Zilina 25:A83–A92. https://doi.org/10.26552/com.C.2023.021

Bin Sulaiman F (2023) Compact city: what is the extent of our exploration for its meanings? A systematic review. Sustainability 15:10302. https://doi.org/10.3390/su151310302

Brand C, Götschi T, Dons E et al (2021) The climate change mitigation impacts of active travel: evidence from a longitudinal panel study in seven European cities. Glob Environ Chang 67:102224. https://doi.org/10.1016/j.gloenvcha.2021.102224

Casarin G, MacLeavy J, Manley D (2023) Rethinking urban utopianism: the fallacy of social mix in the 15-minute city. Urban Stud 60:3167–3186. https://doi.org/10.1177/00420980231169174

Caselli B (2021) From urban planning techniques to 15-minute neighbourhoods. A theoretical framework and GIS-based analysis of Pedestrian accessibility to public services. Eur Transp Eur 1–15. https://doi.org/10.48295/ET.2021.85.10

Coldwell DF, Evans KL (2018) Visits to urban green-space and the countryside associate with different components of mental well-being and are better predictors than perceived or actual local urbanisation intensity. Landsc Urban Plan 175:114–122. https://doi.org/10.1016/j.landurbplan.2018.02.007

Crook L (2021) 15-Minute city concept by Carlos Moreno wins Obel Award 2021. https://www.dezeen.com/2021/10/26/15-minute-city-carlos-moreno-obel-award/

De Falco S (2021) Smart urbanism in historical occurrences and recurrences: An evolutionary perspective between critical and progressive factors. J Urban Regen Renew 14:440. https://doi.org/10.69554/PURY6306

Delmon J (2024) Innovative funding and financing for infrastructure. Cambridge University Press

Djalali A, Gianfrate V, Longo D, Vai E (2020) An advanced design approach to support urban transformations through multi-stakeholder collaborations Strateg Des Res J 12. https://doi.org/10.4013/sdrj.2019.122.09

Elmarakby E, Elkadi H (2024) Prioritising urban heat island mitigation interventions: mapping a heat risk index. Sci Total Environ 948:174927. https://doi.org/10.1016/j.scitotenv.2024.174927

Filazzola A, Shrestha N, MacIvor JS (2019) The contribution of constructed green infrastructure to urban biodiversity: a synthesis and meta-analysis. J Appl Ecol 56:2131–2143. https://doi.org/10.1111/1365-2664.13475

Galiautdinov R (2023) Boosting local economies, pp 55–68

Geyer HS (2024) The theory and praxis of mixed-use development—an integrative literature review. Cities 147:104774. https://doi.org/10.1016/j.cities.2023.104774

Gong J (2024) Strategic analysis on improving the proximity and accessibility in China: an exploratory development based on the study of the 15-minute city. Highlights Sci Eng Technol 86:66–73. https://doi.org/10.54097/jpzasp17

Gössling S (2013) Urban transport transitions: Copenhagen, City of Cyclists. J Transp Geogr 33:196–206. https://doi.org/10.1016/j.jtrangeo.2013.10.013

Ibrahim AHH (2024) Decentralization and its impact on improving public services. Int J Soc Sci 7:45–53. https://doi.org/10.21744/ijss.v7n2.2278

Ingvardson JB, Nielsen OA (2018) Effects of new bus and rail rapid transit systems—an international review. Transp Rev 38:96–116. https://doi.org/10.1080/01441647.2017.1301594

Ivanovic SS, Wood J, Purves RS (2023) Evaluation of bicycle sharing scheme data as a proxy for cycling mobility – How COVID-19 measures influenced cycling in Paris. Transp Res Interdiscip Perspect 22:100937. https://doi.org/10.1016/j.trip.2023.100937

Jin T, Wang K, Xin Y et al (2024) Is a 15-minute city within reach? Measuring multimodal accessibility and carbon footprint in 12 major American cities. Land Use Policy 142:107180. https://doi.org/10.1016/j.landusepol.2024.107180

Jordan J (2013) An Examination of gentrification and related displacement of black residents in Portland's Boise neighborhood, 1990–2010. Portland, OR

Kaplan S, Giacomo Prato C (2015) A spatial analysis of land use and network effects on frequency and severity of cyclist-motorist crashes in the Copenhagen region. Traffic Inj Prev 16:724–731. https://doi.org/10.1080/15389588.2014.1003818

Khavarian-Garmsir AR, Sharifi A, Hajian Hossein Abadi M, Moradi Z (2023a) From garden city to 15-minute city: a historical perspective and critical assessment. Land 12:512. https://doi.org/10.3390/land12020512

Khavarian-Garmsir AR, Sharifi A, Sadeghi A (2023b) The 15-minute city: urban planning and design efforts toward creating sustainable neighborhoods. Cities 132:104101. https://doi.org/10.1016/j.cities.2022.104101

Kohli A (2024) Community engagement and participation in planning of a city. Int J Multidiscip Res 6. https://doi.org/10.36948/ijfmr.2024.v06i03.23793

Krishnan JN, Sebastian J, Joseph JK, et al (2024) Enhanced ridesharing using directions API and geolocation. Int J Res Appl Sci Eng Technol 12:3074–3079. https://doi.org/10.22214/ijraset.2024.62264

Krishnaveni KS, Anilkumar PP (2022) How smart growth can curb urban sprawl: a case study of a rapidly urbanizing city in Kerala, India. Int Arch Photogramm Remote Sens Spat Inf Sci XLIII-B3-2:1355–1360. https://doi.org/10.5194/isprs-archives-XLIII-B3-2022-1355-2022

Kunjir SN, Patil SS, Hingane BS, et al (2023) Managing smart urban transportation with the integration of big data analytic platform. In: 2023 6th International conference on contemporary computing and informatics (IC3I). IEEE, pp 1807–1811

Lee E (2024) Adopting the N-minute city approach to revitalize healthy cities. Acad Urban Heal 3:27–30. https://doi.org/10.59013/auh.2024.3.1.003

Leyden KM (2003) Social capital and the built environment: the importance of walkable neighborhoods. Am J Public Health 93:1546–1551. https://doi.org/10.2105/AJPH.93.9.1546

Li K (2024b) The ideology of sustainable cities. Acad J Manag Soc Sci 6:87–91. https://doi.org/10.54097/yeyah140

Lu M, Diab E (2023) Understanding the determinants of x-minute city policies: a review of the North American and Australian cities' planning documents. J Urban Mobil 3:100040. https://doi.org/10.1016/j.urbmob.2022.100040

Magrinyà F, Mercadé-Aloy J, Ruiz-Apilánez B (2023) Merging green and active transportation infrastructure towards an equitable accessibility to green areas: Barcelona Green Axes. Land 12:919. https://doi.org/10.3390/land12040919

Manaugh K, Badami MG, El-Geneidy AM (2015) Integrating social equity into urban transportation planning: a critical evaluation of equity objectives and measures in transportation plans in North America. Transp Policy 37:167–176. https://doi.org/10.1016/j.tranpol.2014.09.013

Marchigiani E, Bonfantini B (2022) Urban transition and the return of neighbourhood planning. Questioning the proximity syndrome and the 15-minute city. Sustainability 14:5468. https://doi.org/10.3390/su14095468

Martin RJ (2024) Operationalising the 15-minute city with new mobility services. Proc from Annu Transp Conf Aalborg Univ 31. https://doi.org/10.54337/ojs.td.v31i.8706

Mehdipanah R, Marra G, Melis G, Gelormino E (2018) Urban renewal, gentrification and health equity: a realist perspective. Eur J Public Health 28:243–248. https://doi.org/10.1093/eurpub/ckx202

Moos M, Vinodrai T, Revington N, Seasons M (2018) Planning for mixed use: affordable for whom? J Am Plan Assoc 84:7–20. https://doi.org/10.1080/01944363.2017.1406315

Moreno C, Allam Z, Chabaud D et al (2021) Introducing the "15-minute city": sustainability, resilience and place identity in future post-pandemic cities. Smart Cities 4:93–111. https://doi.org/10.3390/smartcities4010006

Morley M, Pafka E (2023) "Density done well" in the pursuit of 20-minute neighbourhoods: navigating fluid discourses in Melbourne. Urban Policy Res 41:148–163. https://doi.org/10.1080/08111146.2023.2198550

Nair R, Miller-Hooks E, Hampshire RC, Bušić A (2013) Large-scale vehicle sharing systems: analysis of Vélib'. Int J Sustain Transp 7:85–106. https://doi.org/10.1080/15568318.2012.660115

Naku DWC (2024) From Urbanism to sustainable urbanism: a new urban agenda for rapidly urbanizing Africa. B P International

Negi H, Jaiswal M, Zainab DK (2024) Local to vocal for local shop-revolutionizing traditional commerce in the digital age. Int J Innov Res Comput Sci Technol (IJIRCST) 307–311

Neutens T (2015) Accessibility, equity and health care: review and research directions for transport geographers. J Transp Geogr 43:14–27. https://doi.org/10.1016/j.jtrangeo.2014.12.006

Newman G, Lee C, Jourdan D et al (2021) Planning the allocation and reallocation of urban space. J Plan Educ Res 41:258–258. https://doi.org/10.1177/0739456X211022680

Omole FO, Olajiga OK, Olatunde TM (2024) Sustainable urban design: a review of eco-friendly building practices and community impact. Eng Sci Technol J 5:1020–1030. https://doi.org/10.51594/estj.v5i3.955

Pardo-Bosch F, Cervera C, Ysa T (2019) Key aspects of building retrofitting: strategizing sustainable cities. J Environ Manage 248:109247. https://doi.org/10.1016/j.jenvman.2019.07.018

Pavate V, Deshmukh VD, Kolekar A, et al (2024) Green building and energy-efficient design. J Environ Eng Stud 9:33–52. https://doi.org/10.46610/JoEES.2024.v09i02.003

Pennington K (2021) Does building new housing cause displacement?: The supply and demand effects of construction in San Francisco. SSRN Electron J. https://doi.org/10.2139/ssrn.3867764

Porter AK, Kohl HW, Pérez A et al (2020) Bikeability: assessing the objectively measured environment in relation to recreation and transportation bicycling. Environ Behav 52:861–894. https://doi.org/10.1177/0013916518825289

Pozoukidou G, Angelidou M (2022) Urban planning in the 15-minute city: revisited under sustainable and smart city developments until 2030. Smart Cities 5:1356–1375. https://doi.org/10.3390/smartcities5040069

Pozoukidou G, Chatziyiannaki Z (2021) 15-Minute city: decomposing the new urban planning Eutopia. Sustainability 13:928. https://doi.org/10.3390/su13020928

Pucher J, Buehler R (2008) Making cycling irresistible: lessons from The Netherlands, Denmark and Germany. Transp Rev 28:495–528. https://doi.org/10.1080/01441640701806612

Pucher J, Dijkstra L (2003) Promoting safe walking and cycling to improve public health: lessons from the Netherlands and Germany. Am J Public Health 93:1509–1516. https://doi.org/10.2105/AJPH.93.9.1509

Rao F, Kong Y, Ng KH et al (2024) Unravelling the Spatial arrangement of the 15-minute city: a comparative study of Shanghai, Melbourne, and Portland. Plan Theory Pract 25:184–206. https://doi.org/10.1080/14649357.2024.2350948

Rigolon A, Derr V, Chawla L (2015) Green grounds for play and learning: an intergenerational model for joint design and use of school and park systems. In: Handbook on green infrastructure. Edward Elgar Publishing

Rossi L, Bianchi G (2024) Sustainable solutions: integrating renewable energy and electric vehicles for cleaner operations. J Energy Res Rev 16:52–63. https://doi.org/10.9734/jenrr/2024/v16i3342

Sallis JF, Cerin E, Conway TL et al (2016) Physical activity in relation to urban environments in 14 cities worldwide: a cross-sectional study. Lancet 387:2207–2217. https://doi.org/10.1016/S0140-6736(15)01284-2

Shilton T, Bauman A, Beger B et al (2024) More people, more active, more often for heart health. Taking Action Phys Activity Glob Heart 19 https://doi.org/10.5334/gh.1308

Shoina M, Voukkali I, Anagnostopoulos A et al (2024) The 15-minute city concept: the case study within a neighbourhood of Thessaloniki. Waste Manag Res J A Sustain Circ Econ 42:694–710. https://doi.org/10.1177/0734242X241259926

Talen E (2024b) New urbanism. In: What cities say. Oxford University Press, New York, pp 82–84

Thaury M-O, Genet S, Maurice L et al (2024) City composition and accessibility statistics in and around Paris Front Big Data 7. https://doi.org/10.3389/fdata.2024.1354007

Thornton LE, Schroers R-D, Lamb KE et al (2022a) Operationalising the 20-minute neighbourhood. Int J Behav Nutr Phys Act 19:15. https://doi.org/10.1186/s12966-021-01243-3

Todd L (2023) Evaluating accessibility for transport planning measuring people's ability to reach desired services and activities. Victoria Transp Policy Inst

van Schalkwyk MCI, Mindell JS (2018) Current issues in the impacts of transport on health. Br Med Bull 125:67–77. https://doi.org/10.1093/bmb/ldx048

Watkins S, Mindell J (2010) Transport and health. Persp Public Health 130:57–57. https://doi.org/10.1177/17579139101300020304

Weller S, van Hulten A (2012) Gentrification and displacement: the effects of a housing crisis on Melbourne's low-income residents. Urban Policy Res 30:25–42. https://doi.org/10.1080/08111146.2011.635410

Xu C, Huang Q, Haase D et al (2024) Cooling effect of green spaces on urban heat island in a Chinese Megacity: increasing coverage versus optimizing spatial distribution. Environ Sci Technol 58:5811–5820. https://doi.org/10.1021/acs.est.3c11048

Yang J (2023) Visualizing and assessing the 15-minute city facility configuration of city region. A study on the Beijing-Tianjin-Hebei urban Agglomeration. Adv Educ Humanit Soc Sci Res 4:63. https://doi.org/10.56028/aehssr.4.1.63.2023

# Time and Space Dimensions 3

## 3.1 Introduction

Urban planning has traditionally concentrated on the spatial design and organization of cities to support rapid population growth, facilitate economic activities, and meet infrastructure demands (Arif and Gupta 2020). The primary focus has been on the efficient allocation of land use for different roles, such as residential, commercial, industrial, and recreational, ensuring that urban growth is resilient, structured, and sustainable (Newman et al. 2021). This conventional approach has often prioritized maximizing space utilization, optimizing traffic flow, and expanding infrastructure networks to accommodate the increasing urbanization.

The 15-minute city concept has revolutionized urban planning by integrating a temporal dimension into the spatial arrangement of urban areas. Time, alongside space, is a pivotal factor in the design, utilization, and experience of urban environments. This approach shifts the focus from merely organizing physical spaces to considering the time it takes for residents to access essential services and amenities. This model aims to foster more resilient, environment-friendly, and equitable cities by redefining the relationship between time and space.

This chapter delves into the intricate dynamics of time and space within the framework of a 15-minute city, exploring their roles, impacts, and challenges across three main areas: urban planning, transportation planning, and the specific design elements of the 15-minute city. The aim was to provide a comprehensive understanding of the importance of temporal and spatial dimensions and how their integration can contribute to more sustainable, liveable, and inclusive urban environments.

## 3.2 Space and Time in Urban Planning

Integrating temporal and spatial dimensions in contemporary urban planning is crucial for creating cities that are both functional and liveable (Levine et al. 2022). These dimensions influence how urban spaces are designed, utilized, and experienced by residents. Spatial dimensions refer to the physical layout and organization of spaces within a city. Effective spatial planning ensures that urban areas are efficiently organized, minimizing travel distances, optimizing land use, and creating coherent neighbourhoods (Bush and Doyon 2019). On the other hand, temporal dimensions address the timing and sequence of urban activities and developments, ensuring that cities can adapt to the dynamic needs of residents throughout the day, week, and year (Hutter and Wiechmann 2022). By integrating both spatial and temporal aspects, urban planners can create



M. Bencekri and S. Lee, *Sustainable Urban Evolution*, Sustainable Development Goals Series,
https://doi.org/10.1007/978-981-96-5146-7_3

**Fig. 3.1** Spatial characteristics of urban planning

environments that are sustainable, resilient, equitable, and responsive to changing patterns of human behaviour and activity, thus improving the functionality of urban spaces (Tao et al. 2019).

### 3.2.1 Spatial Characteristics

Urban planning is a critical discipline that shapes the physical layout and organization of cities and towns. The aim is to create functional, sustainable, and aesthetically pleasing urban environments by considering various spatial characteristics. These characteristics influence how people interact with their surroundings, the efficiency of urban systems, and overall quality of life. This study explores key spatial characteristics (Fig. 3.1) in urban planning, emphasizing their importance and impact.

#### 3.2.1.1 Density and Land Use

Density refers to the concentration of people, buildings, and activities in a specific area. High density can lead to more efficient use of resources and infrastructure, reduce the need for extensive transportation networks, and foster vibrant urban life (Lehmann 2019). For instance, concentrating services near public transport and increasing residential urban density can reduce car dependency, improve travel efficiency, and enhance the quality of life (Ritsema van Eck et al. 2005).

However, high density also poses challenges such as potential overcrowding and strain on public services if not managed correctly (Mitrany 2005). For instance, high urban density can affect social behaviour and attitudes, leading to issues such as stress and mental health problems due to crowded living conditions and lack of personal space. Moreover, it can put strain on public services such as healthcare, education, and transportation, leading to reduced quality and accessibility of these essential services.

Land use planning involves designating areas for residential, commercial, industrial, or recreational purposes (Tian et al. 2017). Mixed-use developments, which combine these uses, are increasingly favoured for their ability to create dynamic and sustainable environments (Evans 2014). For instance, mixed-use development provides various benefits to communities, such as increased walkability, reduced pollution, enhanced local economies, and improved space liveability (Metzinger 2021).

However, mixed-use development faces several significant drawbacks. Regulatory challenges often present the first hurdle: local regulations, zoning laws, and building codes can be rigid and not designed to accommodate the complexities of mixed-use (Metzinger 2021). Compatibility issues further complicate these projects because achieving the right blend of uses within a single development is critical. Incompatible uses, such as residential units that are too close to noisy commercial establishments, can lead to conflicts and reduce the overall value of the project (Ryckewaert et al. 2021). In addition, cultural and contextual differences play a crucial role in the success of mixed-use development (Metzinger 2021). Practices and designs that are successful in one area may not translate well to another because of differing cultural priorities, lifestyles, and local contexts. Securing financing is another issue in mixed-use development. The complexity and perceived risks associated with integrating multiple uses can deter investors and financial institutions, adding another layer of uncertainty to mixed-use development projects Gyourko and Rybczynski 2000). Finally, community opposition can be a significant additional barrier, as concerns about increased density and potential traffic congestion can lead to local resistance (Trounstine 2023). These setbacks highlight the need for careful planning, community engagement, and adaptable strategies to implement mixed-use development successfully.

In summary, density and land use are pivotal in urban planning, shaping city efficiency and liveability. High density can optimize resource use and enhance urban vibrancy; however, if not well planned, it may lead to overcrowding and strain on public services. Effective mixed land use fosters dynamic and sustainable environments, but may face challenges such as regulatory hurdles, compatibility issues, and financing difficulties.

#### 3.2.1.2 Connectivity and Accessibility

Connectivity and accessibility are pivotal concepts in urban planning, integral to creating vibrant, functional, and inclusive cities (Southworth 2016). These elements not only shape the physical layout of urban areas, but also influence economic viability, social inclusion, and overall quality of life (Wittowsky et al. 2019).

Connectivity refers to the ease with which different parts of an urban area can be reached from one another, and refers to how well streets and pathways are linked together (Galpern et al. 2018). High connectivity reduces travel time, enhances the accessibility of services and amenities (Kesarovski and Hernández-Palacio 2023), and promotes economic activity (Ng et al. 2017). Businesses benefit from ease of movement, as employees and customers can reach commercial hubs with minimal delay. Furthermore, well-connected urban areas tend to attract more investment because efficient transport networks reduce logistical challenges and enhance the overall attractiveness of both businesses and residents (Dong et al. 2016).

On the other hand, accessibility ensures that all individuals, regardless of their physical abilities or socioeconomic status, can access essential services and amenities (Knox 1980). Rode and da Cruz (2018) discussed how cities can improve economic development and social well-being by enhancing accessibility, which involves both urban design and transport systems rather than just focusing on mobility and movement. Sánchez Criado and Cereceda Otárola (2016) emphasized the need to include inclusive design and accessibility from the beginning in all urban development projects. Handy and Niemeier (1997) considered accessibility an essential element to reach social inclusion in urban development projects.

#### 3.2.1.3 Public Spaces and Green Areas

Public spaces and green areas are indispensable components of urban environments and play crucial roles in enhancing the quality of urban life.

Public spaces, such as parks, plazas, and streets are essential for fostering social cohesion and community engagement. These areas provide residents with opportunities to interact, relax, and participate in various activities, thereby enhancing the social fabric of the city, as well as the individual well-being of citizens (Sugiyama et al. 2018). The inclusion of seating areas, playgrounds, and pedestrian-friendly walkways in public spaces can transform them into vibrant community hubs (Dong et al. 2019). Archundia (2018), while discussing a case study in Mexico City, emphasized the importance of community involvement, government accountability, and improved pedestrian infrastructure to create more comfortable and safer public spaces.

Green areas, including parks, gardens, and urban forests, play crucial roles in urban sustainability and environmental health (Sugiyama et al. 2018). These spaces help to mitigate the urban heat island effect, reduce air pollution, and provide essential habitats for wildlife (Zhang et al. 2014). Green areas also contribute to residents' mental health by offering spaces for physical activity and reducing stress (Lopes and Camanho 2013).

Cities that prioritize the development and maintenance of public and green spaces often experience a range of social and economic benefits (Ramaiah and Avtar 2019). Attractive public and green areas can increase property value, attract tourists, and boost local businesses. Moreover, they can serve as venues for cultural events, festivals, and markets, fostering a sense of community and belonging among the residents.

Generally, public and green spaces serve as venues for social interactions, recreational and sports activities, and contribute to environmental sustainability by mitigating urban heat islands and enhancing air quality (Arnberger 2012). They foster a sense of community and well-being by offering numerous benefits, including improving mental health by providing residents with areas to socialize, relax, and unwind, and promoting physical health through exercise opportunities (Sugiyama et al. 2018). They also support biodiversity by providing habitats for various species and serving as natural buffers against environmental hazards such as flooding. Additionally, they can increase property values, stimulate local economies by attracting tourists and businesses, and provide educational opportunities in the natural environment. By integrating these spaces into urban planning, cities can create sustainable, resilient, and inclusive environments for all residents.

#### 3.2.1.4 Urban Form and Morphology

Urban form refers to the physical structure and layout of an urban area encompassing its buildings, streets, and open spaces. Urban morphology studies the patterns and processes that shape urban forms over time (Chiaradia 2019). Compact urban forms, characterized by higher density and mixed land use, promote sustainability by reducing the need for extensive transportation networks and preserving natural landscapes.

The concept of a 15-minute city, implemented in Paris, exemplifies the benefits of density and compactness. This model ensures that essential services are readily available within a short walk or bike ride, reducing the need for extensive travel and enhancing urban liveability (Moreno et al. 2021). The 15-minute city model aims to create self-sufficient neighbourhoods where residents can access most of their needs within a 15-minute walk or bike ride.

For example, the compact urban form of Barcelona's Eixample district enhances walkability and accessibility, reduces dependence on cars, and fosters vibrant street life. Conversely, sprawling urban forms, which are often found in many American suburbs, lead to increased car dependency, environmental degradation, and higher infrastructure costs (Camagni et al. 2002). Compact urban forms also help preserve natural landscapes by limiting the spread of urban areas to the surrounding rural regions.

#### 3.2.1.5 Infrastructure and Services

Infrastructure, including transportation systems, utilities, and communication networks, form the backbone of urban areas. Efficient and resilient infrastructure is essential for supporting daily activities and ensuring the smooth functioning of a city (Reiner and McElvaney 2017). Urban planners must consider the spatial distribution of infrastructure and services to ensure equitable access for all residents.

Innovations in infrastructure planning such as the implementation of smart city technologies can enhance the efficiency and sustainability of urban systems (Bifulco et al. 2016). These technologies optimize the functioning of urban systems, making cities more efficient and responsive to residents' needs.

#### 3.2.1.6 Conclusion

Spatial characteristics in urban planning are fundamental to the development of sustainable, liveable, and equitable cities. By carefully considering factors such as density, land use, connectivity, public spaces, urban form, and infrastructure, urban planners can create environments that meet residents' diverse needs. As urbanization continues to present challenges, thoughtful spatial planning will be crucial for promoting high-quality urban life and addressing future needs. Future research and practice should focus on integrating innovative technologies and inclusive design principles to further enhance the sustainability and liveability of urban environments.

### 3.2.2 Temporal Characteristics

Temporal characteristics in urban planning emphasize the integration of time as a crucial element in the design and management of urban spaces (Hutter and Wiechmann 2022). Recognizing the importance of the timing and sequencing of activities, services, and infrastructure can significantly impact the functionality and liveability of cities. Incorporating temporal dimensions into planning makes urban environments more responsive to the dynamic needs of their residents, thereby promoting efficiency, accessibility, and overall quality of life (Munn and Dragićević 2023). Key aspects of the temporal characteristics in urban planning include accessibility and proximity, daily life rhythms, and temporal flexibility (Fig. 3.2).

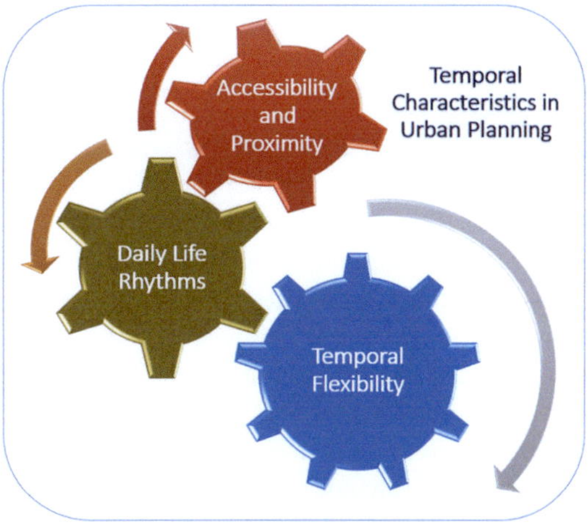


**Fig. 3.2** Temporal characteristics in urban planning

#### 3.2.2.1 Accessibility and Proximity

Temporal planning aims to ensure that essential services, such as healthcare, education, and retail, are accessible within a short timeframe (Neutens et al. 2010). This approach reduces reliance on motorized transport and encourages walking and cycling, contributing to healthier lifestyles and lower social exclusion and environmental impact. This concept is central to the 15-minute city model, which seeks to provide all necessary amenities within a 15-minute walk or bike ride from residents' homes. By strategically locating essential services near residential areas, cities can minimize travel time and enhance the convenience of accessing these services.

For instance, Paris implemented the 15-minute city initiative, leading to the development of multifunctional neighbourhoods where residents can access schools, healthcare facilities, grocery stores, parks, and workplaces within a short distance (Rose et al. 2013). This initiative reduces traffic congestion and pollution, while fostering stronger community ties and a sense of belonging among residents.

#### 3.2.2.2 Daily Life Rhythms

Designing urban spaces that align with the daily routines and needs of residents facilitates seamless transitions between work, home, and leisure activities (Cihanger 2018). Urban planners can optimize the temporal dimension by considering peak activity periods and adjusting the infrastructure and service availability accordingly (Chen et al. 2023). Understanding the temporal patterns of urban life allows the design of spaces that support the ebb and flow of daily activities (Zhang et al. 2019).

#### 3.2.2.3 Temporal Flexibility

Creating adaptable spaces that can serve multiple purposes at different times of the day enhances the efficiency of urban environments (Goessler and Kaluarachchi 2023). Temporal flexibility involves designing buildings and public spaces that can be repurposed to meet varying demands (Redyantanu and Damayanti 2017). This adaptability ensures that urban areas remain vibrant and functional throughout the day and at night.

For example, urban street furniture can incorporate flexible design elements to address the safety, management, and adaptability issues. Siu and Wong (2015) proposed six flexible design principles aimed at enhancing the adaptability and functionality of street furniture and ensuring that they meet the evolving demands of urban spaces.

#### 3.2.2.4 Examples

**Barcelona's Superblocks**

Barcelona implemented the superblock model, reorganizing city blocks to prioritize pedestrian and cyclist movements while restricting car traffic (Palència et al. 2020). This model enhances accessibility by creating car-free zones where essential services are within walking distances. Superblocks also reflect daily life rhythms by providing space for social interaction, play, and relaxation at different times of the day.

**Curitiba's BRT System**

Curitiba's bus rapid transit (BRT) system is an example of an effective temporal planning for public transportation (Lindau et al. 2010). The system includes dedicated bus lanes and synchronized schedules that align with peak commuting times, reduce travel times, and improve the efficiency. Additionally, the city's urban planning integrates residential and commercial areas with transit hubs, ensuring that essential services are easily accessible.

**Flexible Spaces in New York**

New York City has embraced the concept of temporal flexibility through initiatives such as the High Line and the transformation of Times Square (Lang and Rothenberg 2017). The High Line, an elevated park, provides a green space that adapts to different uses throughout the day from quiet morning walks to lively evening events. Times Square was reconfigured to accommodate pedestrian plazas, offering flexible spaces for public gatherings, performances, and leisure activities.

#### 3.2.2.5 Conclusion

Incorporating temporal characteristics into urban planning is essential for creating cities that are efficient, adaptable, and resilient (Ramirez Lopez and Grijalba Castro 2020). Urban planners can design environments that support the dynamic needs of residents by focusing on accessibility and proximity, aligning with daily life rhythms, and fostering temporal flexibility. This holistic approach ensures that urban areas remain vibrant and liveable, providing a high quality of life for all inhabitants. As cities continue to grow and evolve, integrating temporal dimensions into planning will be crucial for addressing contemporary urban challenges and achieving sustainable development goals.

## 3.3 Space and Time in Transportation

Transportation planning plays a pivotal role in the development of sustainable and liveable cities (Wey and Huang 2018). By addressing both spatial and temporal dimensions, urban planners can create transportation systems that align with the 15-minute city model, enhance accessibility, and improve the overall quality of urban life.

### 3.3.1 Spatial Characteristics

Spatial planning in transportation involves the strategic design and placement of infrastructure to support efficient, equitable, and sustainable mobility (Tran and Draeger 2021). The key elements are shown in Fig. 3.3.

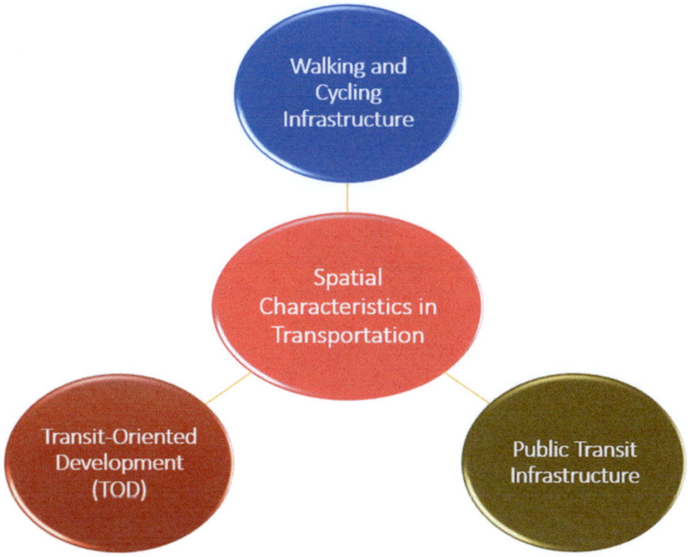


**Fig. 3.3** Spatial characteristics in urban transportation

#### 3.3.1.1 Walking and Cycling Infrastructure

It is essential to develop extensive networks of safe and accessible paths for walking and cycling. Such infrastructure reduces reliance on car travel, decreases traffic congestion, and promotes a healthier lifestyle (Giles-Corti et al. 2016). For example, cities such as Copenhagen and Amsterdam have invested significantly in bike lanes and pedestrian paths, which have not only reduced traffic congestion, but also enhanced public health and lowered carbon emissions (Pucher and Dijkstra 2003). These non-motorized modes of transport are vital for the 15-minute city model, where residents can access essential services within a short walk or bike ride.

#### 3.3.1.2 Public Transit Infrastructure

Ensuring that public transport options are within the easy reach of all residents is crucial for promoting equitable mobility (Welch 2013). This involves strategically placing bus stops, train stations, and other transit facilities near residential, commercial, and recreational areas. Accessible public transport systems provide quick and convenient travel across cities, reducing the need for private car use and lowering GHG emissions (Kinigadner et al. 2020). Additionally, integrating different modes of transport through multimodal hubs enhances the overall accessibility and convenience.

#### 3.3.1.3 Transit-Oriented Development (TOD)

Concentrating urban development around transit hubs maximizes accessibility and reduces travel times (Lyu et al. 2020). TOD involves creating mixed-use neighbourhoods where residential, commercial, and recreational spaces are integrated with public transport networks (Wan et al. 2023). This approach not only enhances the convenience of accessing various services but also stimulates economic activity by increasing foot traffic and business opportunities near transit stations. Cities, such as Hong Kong and Singapore, have successfully implemented TOD, resulting in high public transport usage and reduced dependency on private vehicles (Chow 2014; Niu et al. 2019).

### 3.3.2 Temporal Characteristics

Temporal planning in transportation focuses on optimizing the timing and efficiency of transport systems to minimize travel times and enhance user convenience (Chen et al. 2021). The key strategies (Fig. 3.4) include the following:

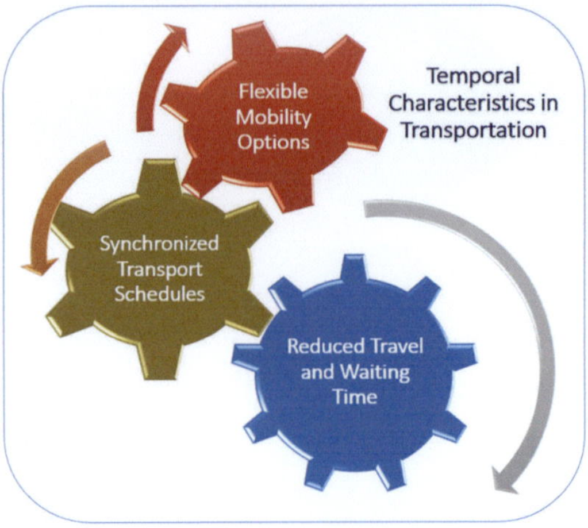


**Fig. 3.4** Temporal characteristics in urban transportation

#### 3.3.2.1 Reduced Travel Times

Designing transport systems that minimize waiting and travel times is essential for ensuring quick access to essential services (Manasra and Toledo 2019). This can be achieved by implementing dedicated bus lanes, expressing transit services, and real-time traffic management systems that prioritize public transport over private vehicles during the peak hours. For instance, the introduction of bus rapid transit (BRT) systems in cities, such as Bogotá and Curitiba, has significantly reduced travel times and improved transit reliability.

#### 3.3.2.2 Synchronized Transport Schedules

Coordinating public transport schedules with residents' daily activities reduces delays and improves convenience (Vlacic 2022). Aligning bus and train schedules with work hours, school timings, and other routine activities ensures that transit services are available when required. This synchronization can significantly enhance user experience by reducing wait times and making public transport a more attractive option. Cities such as Zurich and Munich have implemented highly synchronized transport systems, resulting in efficient and user-friendly public transit networks (Buehler et al. 2019).

#### 3.3.2.3 Flexible Mobility Options

Offering diverse and flexible transportation options caters to various needs at various times (Atasoy et al. 2016). Shared mobility or mobility on-demand services, such as car-sharing, ride-hailing, and bike-sharing, provide alternatives to traditional public transport, accommodating a range of travel preferences and schedules (McKenzie 2020). These options allow for more personalized and efficient travel, particularly for last-mile connectivity and off-peak travel. The proliferation of flexible mobility services in cities, such as San Francisco and Berlin, demonstrates the potential to complement public transit and enhance urban mobility.

#### 3.3.2.4 Conclusion

Integrating spatial and temporal dimensions in transportation planning is crucial for developing cities that are not only functional, but also sustainable and equitable. By focusing on the strategic placement of infrastructure and optimizing transport schedules, urban planners can create environments that support the 15-minute city model, enhance accessibility, and improve residents' overall quality of life. This holistic approach to transportation planning ensures that urban mobility systems are resilient, efficient, and capable of meeting the diverse needs of city residents.

## 3.4 Space and Time in the 15-Minute City

Urban planning has traditionally focused on spatial design to support rapid population growth, facilitate economic activities, and meet infrastructure demands (Arif and Gupta 2020). This conventional approach prioritizes efficient land-use allocation for various purposes, ensuring resilient and sustainable urban growth (Newman et al. 2021). However, the 15-minute city concept has revolutionized urban planning by integrating a temporal dimension, making time a pivotal factor alongside space. This model aims to create cities that

**Fig. 3.5** Spatial characteristics in 15-minute city

are more resilient, environmentally friendly, and equitable by redefining the relationship between time and space. This chapter explores the intricate dynamics of time and space within the framework of a 15-minute city, highlighting their roles, impacts, and challenges.

## 3.4.1 Spatial Characteristics

The spatial characteristics (Fig. 3.5) in the 15-minute city, involve creating a physical layout that supports accessibility, density, and mixed land use (Pozoukidou and Angelidou 2022). These elements are crucial for minimizing travel distances, optimizing land use, and creating coherent neighbourhoods. Effective spatial planning in a 15-minute city ensures that residents can access most, if not all, of their daily needs within a short walk or bike ride from their homes (Zhang et al. 2023). This approach not only fosters vibrant and sustainable urban environments but also reduces car dependency and enhances the efficiency of urban systems (Lehmann 2019).

### 3.4.1.1 Accessibility and Proximity

The principles of accessibility and proximity form the foundation of the 15-minute city model, emphasizing a human-centric approach to urban planning. Proximity ensures that essential services are located within a 15-minute walk or bike ride, significantly reducing travel times and fostering active transportation (Allam et al. 2022a). By ensuring services are evenly distributed across the city, this model prevents the overconcentration of amenities in specific areas, thereby promoting balanced urban development. Additionally, proximity supports environmental sustainability by minimizing transportation-related greenhouse gas (GHG) emissions, which has far-reaching benefits for urban air quality and public health.

Accessibility complements proximity by focusing on equitable access to key services, such as healthcare, education, grocery stores, and recreational facilities, for all residents (Zhang et al. 2023). Strategic placement of these amenities reduces reliance on long commutes and encourages the adoption of sustainable transportation modes like walking and cycling. Enhanced accessibility not only improves convenience and quality of life but also promotes social equity by ensuring that residents, regardless of socioeconomic status, can easily reach the services they need (Neutens 2015).

Barcelona's implementation of the 15-minute city initiative provides a compelling example of this model in action. By creating multi-functional neighbourhoods where residents can access essential amenities with ease, the city has reduced traffic congestion, improved air quality, and fostered stronger community ties. However, research by (Graells-Garrido et al. 2021) highlights that the distribution of these benefits varies across neighbourhoods, often reflecting pre-existing socioeconomic inequalities. This underscores the importance of localized, context-sensitive planning to ensure that the 15-minute city framework delivers equitable outcomes for all urban residents.

By integrating proximity and accessibility, the 15-minute city model offers a transformative vision for sustainable urban and transportation development. However, its success depends on addressing variations in neighbourhood needs and ensuring that planning processes prioritize inclusivity and equity.

#### 3.4.1.2 Dense and Compact Development

Dense and compact development is a fundamental spatial characteristic of the 15-minute city model, enabling efficient land use and enhancing urban accessibility. High-density urban environments concentrate amenities and services within smaller areas, ensuring they are within walking or cycling distance for residents (Lehmann 2016). This approach reduces travel demand, making cities more pedestrian-friendly and decreasing reliance on private vehicles, which in turn supports lower greenhouse gas (GHG) emissions and improves air quality.

Compact development also underpins the success of public transportation systems by creating a critical mass of users within proximity to transit hubs. This enhances system efficiency, lowers operational costs, and provides residents with reliable alternatives to car travel. Furthermore, by limiting urban sprawl, compact development helps preserve surrounding natural and rural areas, protecting ecosystems and promoting biodiversity (Duranton and Puga 2020).

The integration of mixed-use zoning plays a pivotal role in compact development, allowing residential, commercial, and recreational spaces to coexist. This zoning strategy fosters vibrant, multifunctional neighbourhoods where daily needs can be met within a short distance. It also promotes social interaction and community cohesion by creating shared spaces and reducing the isolation often associated with low-density suburban designs.

Cities like Tokyo and Paris exemplify the principles of high-density, compact urban planning. In Tokyo, dense residential and commercial zones are seamlessly connected by an extensive and efficient public transit network, reducing reliance on cars and supporting active modes of transportation. Similarly, Paris has embraced the 15-minute city concept by redesigning its neighbourhoods to prioritize pedestrian and cycling infrastructure, creating walkable communities where essential services are easily accessible. Both cities show how thoughtful planning can balance density with liability, reducing traffic congestion and environmental impact while enhancing quality of life.

However, while dense development offers numerous benefits, it must be carefully managed to avoid challenges such as overcrowding, lack of green spaces, or reduced affordability. Planners must adopt a mixed and optimized land-use and zoning, incorporating sufficient public parks, affordable housing, and infrastructure upgrades to balance sustainable and resilience goals with the needs of diverse populations.

#### 3.4.1.3 Mixed and Optimized Land Use

Mixed and optimized land use are pivotal spatial characteristics of the 15-minute city model.

##### Mixed Land Use

Mixed land use involves integrating residential, commercial, and recreational functions within the same area, allowing residents to live, work, and access essential services without traveling long distances. This approach fosters dynamic neighbourhoods that support local economic activity, provide opportunities for social interaction, and reduce dependency on private vehicles (Ferm and Jones 2016). By bringing people closer to amenities, mixed-use development also encourages sustainable mobility options, such as walking and cycling.

Cities like London and New York illustrate the success of mixed-use planning. For example, neighbourhoods in these cities seamlessly combine housing, offices, retail spaces, and public amenities, creating vibrant urban centres that reduce commute times and promote an active urban lifestyle. These mixed-use areas not only enhance convenience but also contribute to economic resilience by supporting small businesses and encouraging foot traffic.

#### Optimized Land Use

Optimized land use focuses on maximizing the efficiency of available land to address urban challenges while preserving environmental and social equity. This involves strategic zoning that integrates green spaces, public parks, and recreational areas into urban planning to support residents' physical and mental well-being. Additionally, land-use optimization prioritizes the redevelopment of underutilized areas to curb urban sprawl and protect surrounding natural landscapes (Ligmann-Zielinska et al. 2008).

A leading example of land-use optimization is Barcelona's superblocks. These innovative, car-free zones repurpose existing streetscapes to prioritize pedestrians and cyclists, enhance accessibility, and reduce vehicular pollution. Research has shown that superblocks improve air quality, increase public space availability, and foster stronger community engagement (Vardakas et al. 2018); (Scudellari et al. 2020).

#### Mixed and Optimized Land Use Synergy

When combined, mixed and optimized land use create neighbourhoods that balance functionality, accessibility, and sustainability. Mixed-use zoning ensures that services and amenities are conveniently located, while optimized land use guarantees that these spaces are designed for efficiency and inclusivity. Together, they reduce travel demand, lower greenhouse gas emissions, and foster equitable access to essential services.

The integration of mixed and optimized land use in the 15-minute city model transforms urban areas into accessible and sustainable spaces. For instance, the integration of mixed-use planning with land-use optimization can create multifunctional areas that support diverse demographics and reduce socioeconomic disparities. However, careful planning is essential to mitigate challenges, such as overcrowding, gentrification, or conflicting land-use priorities.

#### 3.4.1.4 Public and Green Spaces

Public and green spaces such as parks, plazas, and community centres, play a vital role in 15-minute cities (Zong 2023). These spaces provide areas for social interaction, recreation, and cultural activities, contributing to the overall liveability and attractiveness of urban neighbourhoods. Well-designed public spaces enhance the aesthetic appeal of a city and promote the mental and physical well-being of residents (Salam 2020). Cities such as Melbourne have successfully integrated extensive public spaces within their urban planning strategies, supporting the 15-minute city model and fostering a sense of community (Bush et al. 2020).

#### 3.4.1.5 Sustainable Infrastructure

Sustainable infrastructure supports the goals of a 15-minute city by providing environmentally friendly transportation options, energy-efficient buildings, and resilient urban systems (Allam et al. 2022a). This includes the development of bike lanes, pedestrian pathways, and efficient public transit networks that reduce the reliance on cars. Additionally, incorporating green building practices and renewable energy sources into urban development helps reduce environmental impacts and promotes long-term sustainability. For example, Amsterdam has implemented extensive bike infrastructure and energy-efficient buildings, making it a model for sustainable urban living (Dobbelsteen et al. 2019).

### 3.4.2 Temporal Characteristics

The 15-minute city is fundamentally shaped by temporal factors that influence how residents experience urban life. These factors, depicted in Fig. 3.6, emphasize the need for urban environments to align spatial planning with time-related dynamics, ensuring accessibility, inclusivity, and vibrancy.

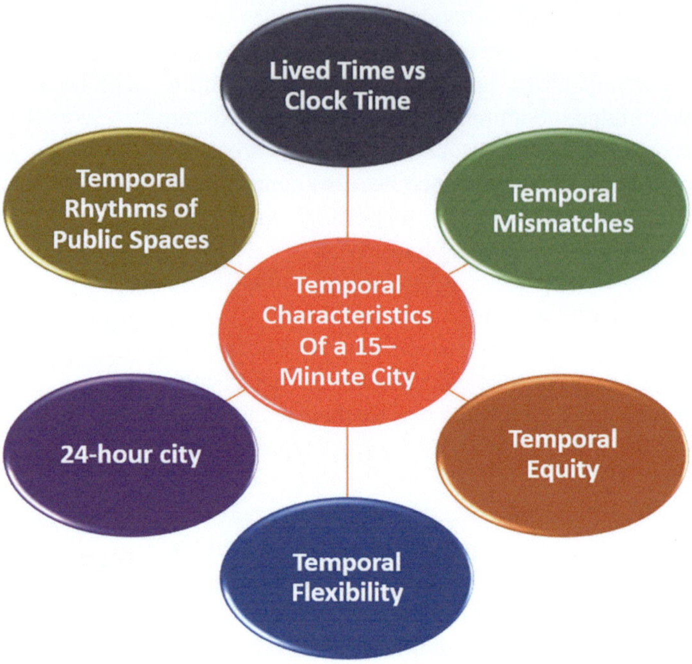


**Fig. 3.6** Temporal characteristics of 15-minute city

#### 3.4.2.1 Lived Time Versus Clock Time

Urban environments are shaped by two interconnected temporal dimensions: ***lived time*** and ***clock time***. Lived time reflects the subjective experience of time, shaped by personal routines, social interactions, and cultural contexts. It represents how individuals perceive and navigate their daily lives. Conversely, clock time is the standardized, objective measurement that structures societal activities, from work schedules to public transportation timetables (Koenderink et al. 2020); (Pozoukidou and Angelidou 2022).

Balancing these temporal dimensions is crucial in the design of 15-minute cities, where the goal is to ensure accessibility and inclusivity within a short distance. For instance, urban planners can design spaces that support flexible schedules, such as libraries open late for evening visitors or multi-use public spaces that accommodate different activities throughout the day.

This balance ensures that urban environments meet both individual needs and collective expectations. Recognizing the interplay between lived and clock time allows planners to create cities that align with diverse lifestyles while maintaining the efficiency and functionality required for cohesive urban life.

#### 3.4.2.2 Temporal Mismatches

Temporal mismatches arise when the timing of urban services does not align with residents' schedules, creating barriers to accessibility and reducing the efficiency of these services. For instance, public transportation systems that fail to accommodate peak commuting hours can lead to overcrowding at certain times and underutilization during others. Similar conflicts occur in healthcare services with limited hours or recreational facilities that close early. Addressing these mismatches requires careful coordination and planning to ensure that urban services are available when residents need them the most (Ruiz and Timmermans 2006).

In the context of a 15-minute city, where the goal is to provide timely and convenient access to services within a short distance, resolving temporal mismatches is crucial. Planners can address these issues through measures such as synchronizing transit schedules with local activity patterns, extending the operational hours of essential services, and introducing flexible service models that respond dynamically to demand.

By minimizing temporal mismatches, cities can better cater to the diverse lifestyles of their residents, ensuring that services are available when they are most needed. This alignment not only improves the user experience but also promotes equity by reducing the temporal barriers that disproportionately affect those with non-traditional schedules, such as shift workers or caregivers.

#### 3.4.2.3 Temporal Equity

Temporal equity ensures that all residents, regardless of their schedules or time constraints, have fair access to essential services and opportunities. This principle is especially critical for supporting vulnerable populations, such as shift workers, caregivers, and individuals with non-traditional work hours, who often face challenges in accessing essential services during standard operating hours due to limited-service availability (Neutens et al. 2010).

In the context of a 15-minute city, temporal equity plays a vital role in fostering inclusivity. For example, ensuring 24-hour access to critical services like healthcare, emergency response, and transportation is essential to meet the needs of diverse communities. Public libraries, recreational facilities, and grocery stores with extended or flexible hours can further enhance accessibility, reducing disparities in service availability across different timeframes.

Achieving temporal equity requires careful urban planning and policymaking. Cities can adopt measures such as staggered service hours, on-demand transportation options, and digital platforms that enable remote access to services. These strategies ensure that residents, regardless of their schedules, can fully participate in urban life, thereby promoting social equity and inclusivity.

Temporal equity is a cornerstone of inclusive urban planning, ensuring that services are available across diverse timeframes to accommodate the varying lifestyles of the residents. The vulnerable populations, such as shift workers and those with non-traditional work hours, can have across to all essential services according to their adequate timeframes (Pozoukidou and Angelidou 2022).

#### 3.4.2.4 Temporal Flexibility

Temporal flexibility refers to the capacity of urban spaces to adapt to different uses throughout the day, ensuring that they remain functional, vibrant, and inclusive. This concept highlights the importance of designing buildings and public spaces that can accommodate diverse time-related demands and activities, thereby maximizing their efficiency and usability (Cialdea Mayer et al. 2016).

For instance, an office building might transition into a community centre or an event space after working hours, while public plazas can serve as markets during the day and host cultural performances or social gatherings in the evening. These adaptive uses keep urban spaces relevant and engaging, fostering continuous activity throughout the day and night. Such designs also enhance community interaction by providing opportunities for a variety of demographics to engage with the same space at different times.

In the context of the 15-minute city, where accessibility and resource efficiency are central goals, temporal flexibility is particularly significant. Flexible temporal urban designs, such as modular furniture and multipurpose spaces, can accommodate a variety of activities during a variety of flexible time frames, fostering a dynamic and efficient urban fabric (Goessler and Kaluarachchi 2023). For example:

- **Modular and Reconfigurable Designs**: Buildings with movable walls or flexible layouts can transition from office spaces during the day to recreational or social hubs in the evening.
- **Shared Public Spaces**: Parks or plazas can accommodate morning fitness classes, midday markets, and evening concerts through thoughtful programming and infrastructure.
- **Smart Technologies**: Dynamic lighting systems, automated scheduling platforms, and responsive furniture systems enable spaces to adapt seamlessly to changing demands.

Temporal flexibility not only optimizes urban functionality but also contributes to sustainability. By reusing and repurposing spaces rather than building additional infrastructure, cities can reduce resource consumption and minimize their environmental impact.

Moreover, this approach enhances inclusivity by addressing the diverse schedules and needs of urban residents. For example, shift workers, caregivers, and students can benefit from spaces that operate or transition outside traditional business hours, ensuring they have equal opportunities to access services and amenities.

Embedding temporal flexibility into urban planning, allows the creation of environments that evolve dynamically with the rhythms of daily life. This adaptability is key to achieving the 15-minute city's broader objectives of accessibility, inclusivity, and sustainability, while also fostering vibrant and resilient urban communities (Cialdea Mayer et al. 2016).

#### 3.4.2.5 The 24-Hour City

The 24-hour city is an urban model that supports continuous activity and operations throughout the day and night. It offers numerous benefits, particularly in fostering vibrant, inclusive, and economically dynamic urban environments. However, the concept also presents challenges that require careful management to ensure sustainability and equity (Smith and Hall 2013).

##### Opportunities

A 24-hour city can significantly boost economic activity by extending business hours, creating new employment opportunities, and promoting tourism. For example, late-night retail, dining, and entertainment options can cater to individuals with non-traditional schedules, such as shift workers and tourists, while creating a vibrant urban culture that fosters social interaction and community engagement (Chipungu 2021). This round-the-clock accessibility can also make cities more attractive for businesses and residents alike, reinforcing their global competitiveness.

##### Challenges

Despite its benefits, the 24-hour city concept poses several challenges. Continuous operations can lead to:

- **Increased Noise Pollution**: Late-night activities may disturb residential neighbourhoods, requiring noise control measures.
- **Higher Energy Consumption**: Extended operating hours for businesses, public spaces, and transportation systems can strain energy resources, potentially increasing carbon emissions.
- **Infrastructure Demands**: Maintaining reliable 24-hour public transportation, lighting, and safety services requires significant investment and robust planning.
- **Social Inequities**: If not implemented thoughtfully, the 24-hour city might primarily cater to certain groups (e.g., affluent residents, tourists) while excluding others, such as lower-income populations who may not benefit equally from extended services.

##### Relevance to the 15-Minute City

While the 24-hour city and 15-minute city are distinct concepts, they share common goals of accessibility and inclusivity. In a 15-minute city, the integration of 24-hour services, such as healthcare, public transportation, and essential retail, ensures that residents can access their needs regardless of the time. For example, 24-hour grocery stores or transportation options can complement the 15-minute city model by supporting those who work irregular hours or require flexibility in accessing services.

##### Addressing Challenges

To balance the opportunities and challenges of a 24-hour city, urban planners can adopt innovative solutions, such as:

- **Zoning Policies**: Strategically locating nightlife districts or late-night services away from residential areas can reduce noise-related conflicts.

- **Energy-Efficient Systems**: Investing in sustainable technologies, such as LED streetlights and renewable energy sources, can mitigate the environmental impact of 24-hour operations.
- **Equitable Service Design**: Ensuring that extended-hour services cater to diverse populations—such as shift workers, caregivers, and students—can promote inclusivity and address potential social inequities.
- **Community Engagement**: Involving residents in the planning process can help identify and prioritize services that meet local needs while minimizing disruptions.

Incorporating 24-hour city elements into urban planning can create dynamic, resilient cities that operate efficiently and inclusively around the clock. When integrated thoughtfully with the principles of the 15-minute city, the 24-hour city model can enhance accessibility and improve the quality of life for all residents while addressing the social and environmental challenges of continuous urban activity.

#### 3.4.2.6 Temporal Rhythms of Public Spaces

Public spaces are deeply influenced by the temporal rhythms of urban life, reflecting the dynamic ways in which they are utilized at different times of the day. These rhythms shape how parks, plazas, and recreational areas support various activities, ranging from morning exercise to midday leisure and evening social gatherings. Rather than being static, public spaces must evolve throughout the day to meet the shifting needs of their communities (Nemeškal et al. 2020; Gibert-Flutre 2022).

Designing public spaces to accommodate these temporal patterns is essential for ensuring their functionality, efficiency, flexibility, and relevance (Gibert-Flutre 2022). In the context of the 15-minute city model, temporal rhythms are especially critical for ensuring that public spaces remain accessible and vibrant throughout the day. The 15-minute city prioritizes proximity and equitable access, but integrating time-based adaptability into public spaces ensures that residents can find value in these areas regardless of their schedules. For instance, parks designed with jogging paths and shaded seating can serve early-morning exercisers, while the same areas may support family picnics or small community gatherings during the day, and outdoor movie nights or performances in the evening.

Planning for temporal rhythms also addresses challenges like fluctuating demand during peak and off-peak times. Public spaces designed to evolve with the rhythms of urban life ensure that areas are neither overcrowded during specific times nor underutilized during others. For example, a neighbourhood plaza in a 15-minute city can be programmed for morning farmers' markets, lunchtime seating for nearby workers, and evening cultural events, creating a dynamic and multifunctional urban environment.

Integrating time-aware designs ensures that public spaces meet the changing needs of residents, fostering stronger community connections and enhancing the liveability of cities, making the 15-minute city framework more effective.

## 3.5 Temporal-Spatial Optimization

Temporal-spatial optimization refers to the strategic integration of time-sensitive planning and spatial organization to enhance the efficiency, adaptability, and inclusivity of urban environments. By leveraging technology and thoughtful design, cities can better align services, infrastructure, and public spaces with the temporal needs of residents. This approach is particularly vital to the 15-minute city, where the goal is to ensure proximity and accessibility while addressing the diverse rhythms of urban life.

### 3.5.1 Optimization Characteristics

Temporal-spatial optimization involves combining smart technologies and adaptive urban design to create systems that respond dynamically to temporal and spatial demands. Technologies like real-time traffic management, dynamic public transport scheduling, and digital service platforms enable cities to adjust based on changing conditions, improving service delivery and resource allocation (Ismagilova et al. 2019).

For example, a smart public transit system could adjust bus routes and schedules in real time based on demand, reducing waiting times and increasing efficiency. Similarly, public spaces such as parks or plazas can integrate modular infrastructure to transition seamlessly from morning markets to evening performances, aligning with the temporal rhythms of urban life.

In the 15-minute city, this optimization ensures not only spatial proximity to essential services but also temporal alignment. For instance, extending the hours of healthcare facilities, libraries, or recreational centres accommodates residents with non-traditional schedules, such as shift workers or caregivers, ensuring greater inclusivity.

### 3.5.2 Technology: A Tool, not a Goal

While technology is central to temporal-spatial optimization, it should remain a means to an end rather than the goal of urban planning. The primary focus of urban design must be on creating liveable, inclusive, and sustainable cities (Cook and Karvonen 2024). Technology should support broader planning objectives rather than dictate them, ensuring that human-centred needs remain the priority.

For example, while artificial intelligence and data analytics can streamline urban services or optimize land use, planners must integrate these tools with equitable design principles. This means addressing both physical and temporal accessibility, ensuring services are within reach and available at the right times for all demographics.

In the context of the 15-minute city, technology plays a supporting role by facilitating real-time wayfinding systems, shared mobility platforms, and adaptive energy grids that empower residents. However, urban planners must remain vigilant to avoid over-reliance on technology and ensure that solutions enhance, rather than replace, community-driven design.

### 3.5.3 Opportunities and Challenges

The integration of technology into urban planning offers numerous opportunities, including improved service delivery, enhanced data collection, and more efficient resource management (Ismagilova et al. 2019).

#### 3.5.3.1 Opportunities

Temporal-spatial optimization offers several benefits (Ismagilova et al. 2019)), which perfectly align with the principles of the 15-minute city:

- **Efficient Resource Management**: Technologies like smart grids and water monitoring systems can optimize resource allocation, reducing waste and improving sustainability.
- **Enhanced Public Space Usage**: Modular designs and programmable infrastructure can transform public spaces to accommodate diverse activities, increasing their utility and appeal.
- **Dynamic Transportation Systems**: Real-time transit solutions, such as adaptive bus routes and bike-sharing programs, enhance mobility while reducing traffic congestion and emissions.
- **Data-Driven Planning**: Analytics tools provide insights into mobility patterns, service usage, and community needs, enabling evidence-based decision-making.

#### 3.5.3.2 Challenges

However, implementing temporal-spatial optimization comes with obstacles that must be addressed (Pozoukidou and Angelidou 2022):

- **Privacy Concerns**: The collection of real-time data, while essential for optimization, raises issues around surveillance and misuse of personal information. Transparent data governance and strict privacy policies are crucial to address these concerns.

- **Digital Inequities**: Not all residents have equal access to digital technologies, creating a digital divide that may exclude vulnerable populations from benefiting fully from smart systems. Planners must ensure equitable access to technological tools and services.
- **High Costs**: Implementing smart infrastructure and maintaining it over time requires significant financial investment, as well as training for urban planners and city workers.
- **Over-Reliance on Technology**: Cities must avoid becoming overly dependent on technological systems, which are vulnerable to outages, cybersecurity threats, or obsolescence. Planners should incorporate redundancies and prioritize robust, low-tech solutions where possible.

#### 3.5.3.3 Relevance in the 15-Minute City

Temporal-spatial optimization is a vital enabler of the 15-minute city, ensuring that services and amenities are not only physically accessible but also temporally aligned with residents' diverse lifestyles. For example:

- **Flexible Public Spaces**: A central plaza could host morning fitness programs, lunchtime dining, and evening performances, ensuring continuous engagement throughout the day.
- **Dynamic Mobility Solutions**: Shared mobility systems like bike-sharing or e-scooters can be repositioned in real time to meet peak demand in specific neighbourhoods.
- **Extended Service Hours**: Facilities like healthcare clinics, grocery stores, or libraries can operate on staggered schedules to accommodate residents with non-traditional work hours.

By integrating temporal-spatial optimization into urban design, cities can not only enhance accessibility and liveability but also foster resilience in the face of changing urban dynamics. This approach supports the 15-minute city's broader goals of proximity, sustainability, and inclusivity, ensuring that all residents benefit from an adaptable and responsive urban environment.

## 3.6 Conclusion

The integration of **temporal and spatial dimensions** into urban planning represents a critical shift toward creating cities that are efficient, inclusive, and adaptable. By combining spatial strategies, such as high-density development, mixed land use, and connectivity, with temporal considerations like flexibility, daily life rhythms, and equity, planners can design urban environments that meet the dynamic and diverse needs of their residents. This holistic approach ensures that cities remain vibrant, functional, and sustainable in the face of evolving urban challenges.

The **15-minute city model** exemplifies this integration by prioritizing accessibility and proximity. Through compact development and mixed-use zoning, cities can reduce travel distances, promote active transportation, and minimize car dependency, supporting environmental sustainability and urban liveability (Lehmann 2019). Public spaces and green areas further enhance the social and environmental fabric of cities, fostering mental well-being, social interaction, and biodiversity (Sugiyama et al. 2018). Meanwhile, sustainable infrastructure supports greener lifestyles, aligning with global sustainability goals.

Temporal characteristics complement these spatial strategies by ensuring that urban systems respond dynamically to the rhythms of daily life. Concepts such as **temporal equity** ensure that all residents, regardless of their schedules or socioeconomic status, have access to essential services. **Temporal flexibility**, on the other hand, enables urban spaces to adapt to varying uses throughout the day, ensuring continuous functionality and inclusivity. Technologies like real-time transportation management and modular design further enhance **temporal-spatial optimization**, allowing cities to efficiently address the changing needs of their communities (Pozoukidou and Angelidou 2022).

The success of this integrated framework is evident in cities like **Barcelona**, with its superblocks, and **Copenhagen**, with its extensive cycling infrastructure Pozoukidou and Angelidou (2022). These examples demonstrate how thoughtful planning can transform urban areas into sustainable, equitable, and vibrant spaces. However, the implementation of the 15-minute city framework is not without challenges. Addressing privacy concerns in technology, overcoming digital divides, and ensuring that compact development does not lead to overcrowding or gentrification will require thoughtful policies, robust community engagement, and careful balancing of priorities.

In conclusion, the integration of temporal and spatial dimensions provides a transformative framework for urban planning, ensuring that cities are not only functional but also responsive, inclusive, and sustainable. By adopting the principles of the 15-minute city, urban planners can create communities where residents can live, work, and thrive within a short distance of their homes, significantly enhancing quality of life. Moving forward, urban planning must prioritize innovation, equity, and adaptability to meet the growing complexities of urbanization while ensuring that the cities of tomorrow remain resilient and supportive for all.

## References

Allam Z, Bibri SE, Chabaud D, Moreno C (2022a) The theoretical, practical, and technological foundations of the 15-minute city model: proximity and its environmental, social and economic benefits for sustainability. Energies 15:6042. https://doi.org/10.3390/en15166042

Archundia PEC (2018) PW 0994 Healthy public space as goal to safe communities in mexico city. In: Abstracts. BMJ Publishing Group Ltd., pp A199.2–A199

Arif M, Gupta K (2020) Spatial development planning in peri-urban space of Burdwan City, West Bengal, India: statutory infrastructure as mediating factors. SN Appl Sci 2:1779. https://doi.org/10.1007/s42452-020-03587-0

Arnberger A (2012) Urban densification and recreational quality of public urban green spaces—a Viennese case study. Sustainability 4:703–720. https://doi.org/10.3390/su4040703

Atasoy B, Ikeda T, Ben-Akiva ME (2016) Optimizing a flexible mobility on demand system. Transp Res Rec J Transp Res Board 2563:76–85. https://doi.org/10.3141/2536-10

Bifulco F, Tregua M, Amitrano CC, D'Auria A (2016) ICT and sustainability in smart cities management. Int J Public Sect Manag 29:132–147. https://doi.org/10.1108/IJPSM-07-2015-0132

Buehler R, Pucher J, Dümmler O (2019) Verkehrsverbund: the evolution and spread of fully integrated regional public transport in Germany, Austria, and Switzerland. Int J Sustain Transp 13:36–50. https://doi.org/10.1080/15568318.2018.1431821

Bush J, Doyon A (2019) Building urban resilience with nature-based solutions: how can urban planning contribute? Cities 95:102483. https://doi.org/10.1016/j.cities.2019.102483

Bush J, Coffey B, Fastenrath S (2020) Governing urban greening at a metropolitan scale: an analysis of the Living Melbourne strategy. Aust Plan 56:95–102. https://doi.org/10.1080/07293682.2020.1739093

Camagni R, Gibelli MC, Rigamonti P (2002) Urban mobility and urban form: the social and environmental costs of different patterns of urban expansion. Ecol Econ 40:199–216. https://doi.org/10.1016/S0921-8009(01)00254-3

Chen X, Wang Y, Ma X (2021) Integrated optimization for commuting customized bus stop planning, routing design, and timetable development with passenger spatial-temporal accessibility. IEEE Trans Intell Transp Syst 22:2060–2075. https://doi.org/10.1109/TITS.2020.3048520

Chen H, Li J, Wang Y et al (2023) Evaluating trade-offs in ecosystem services for blue–green–grey infrastructure planning. Sustainability 16:203. https://doi.org/10.3390/su16010203

Chiaradia AJF (2019) Urban morphology/urban form. In: The Wiley blackwell encyclopedia of urban and regional studies. Wiley, pp 1–6

Chipungu L (2021) Diversifying inclusivity through the night economy: an observation from the terraces of Durban and Cape Town, pp 133–161

Chow ASY (2014) Urban design, transport sustainability and residents' perceived sustainability: a case study of transit-oriented development in Hong Kong. J Comp Asian Dev 13:73–104. https://doi.org/10.1080/15339114.2014.892818

Cialdea Mayer M, Orlandini A, Umbrico A (2016) Planning and execution with flexible timelines: a formal account. Acta Inform 53:649–680. https://doi.org/10.1007/s00236-015-0252-z

Cihanger D (2018) Spaces by people: an urban design approach to everyday life. METU J Fac Archit. https://doi.org/10.4305/METU.JFA.2018.2.2

Cook M, Karvonen A (2024) Urban planning and the knowledge politics of the smart city. Urban Stud 61:370–382. https://doi.org/10.1177/00420980231177688

Dong L, Li R, Zhang J, Di Z (2016) Population-weighted efficiency in transportation networks. Sci Rep 6:26377. https://doi.org/10.1038/srep26377

Dong W, Cao X, Wu X, Dong Y (2019) Examining pedestrian satisfaction in gated and open communities: an integration of gradient boosting decision trees and impact-asymmetry analysis. Landsc Urban Plan 185:246–257. https://doi.org/10.1016/j.landurbplan.2019.02.012

Duranton G, Puga D (2020) The economics of urban density. Cambridge, MA

Evans G (2014) Living in the city. In: Wellbeing. Wiley, pp 1–28

Ferm J, Jones E (2016) Mixed-use 'regeneration' of employment land in the post-industrial city: challenges and realities in London. Eur Plan Stud 24:1913–1936. https://doi.org/10.1080/09654313.2016.1209465

Galpern P, Ladle A, Alaniz Uribe F et al (2018) Assessing urban connectivity using volunteered mobile phone GPS locations. Appl Geogr 93:37–46. https://doi.org/10.1016/j.apgeog.2018.02.009

Gibert-Flutre M (2022) Rhythmanalysis: Rethinking the politics of everyday negotiations in ordinary public spaces. Environ Plan C Polit Sp 40:279–297. https://doi.org/10.1177/23996544211020014

Giles-Corti B, Vernez-Moudon A, Reis R et al (2016) City planning and population health: a global challenge. Lancet 388:2912–2924. https://doi.org/10.1016/S0140-6736(16)30066-6

Goessler T, Kaluarachchi Y (2023) Smart adaptive homes and their potential to improve space efficiency and personalisation. Buildings 13:1132. https://doi.org/10.3390/buildings13051132

Graells-Garrido E, Serra-Burriel F, Rowe F et al (2021) A city of cities: measuring how 15-minutes urban accessibility shapes human mobility in Barcelona. PLoS ONE 16:e0250080. https://doi.org/10.1371/journal.pone.0250080

Gyourko JE, Rybczynski W (2000) Financing new urbanism projects: obstacles and solutions. Hous Policy Debate 11:733–750. https://doi.org/10.1080/10511482.2000.9521384

Handy SL, Niemeier DA (1997) Measuring accessibility: an exploration of issues and alternatives. Environ Plan A Econ Sp 29:1175–1194. https://doi.org/10.1068/a291175

Hutter G, Wiechmann T (2022) Time, temporality, and planning—comments on the state of art in strategic spatial planning research. Plan Theory Pract 23:157–164. https://doi.org/10.1080/14649357.2021.2008172

Ismagilova E, Hughes L, Dwivedi YK, Raman KR (2019) Smart cities: advances in research—an information systems perspective. Int J Inf Manage 47:88–100. https://doi.org/10.1016/j.ijinfomgt.2019.01.004

Kesarovski T, Hernández-Palacio F (2023) Time, the other dimension of urban form: measuring the relationship between urban density and accessibility to grocery shops in the 10-minute city. Environ Plan B Urban Anal City Sci 50:44–59. https://doi.org/10.1177/23998083221103259

Kinigadner J, Büttner B, Wulfhorst G, Vale D (2020) Planning for low carbon mobility: Impacts of transport interventions and location on carbon-based accessibility. J Transp Geogr 87:102797. https://doi.org/10.1016/j.jtrangeo.2020.102797

Knox PL (1980) Measures of accessibility as social indicators: a note. Soc Indic Res 7:367–377. https://doi.org/10.1007/BF00305607

Koenderink J, Pinna B, van Doorn A (2020) Capricious texture of time in awareness and art. Art Percept 8:188–236. https://doi.org/10.1163/22134913-bja10008

Lang S, Rothenberg J (2017) Neoliberal urbanism, public space, and the greening of the growth machine: New York City's High Line park. Environ Plan A Econ Sp 49:1743–1761. https://doi.org/10.1177/0308518X16677969

Lehmann S (2016) Sustainable urbanism: towards a framework for quality and optimal density? Futur Cities Environ 2:8. https://doi.org/10.1186/s40984-016-0021-3

Lehmann S (2019) Understanding the benefits of urban density. Urban regeneration. Springer International Publishing, Cham, pp 79–107

Levine D, Sussman S, Aharon-Gutman M (2022) Spatial-temporal patterns of self-organization: adynamic 4D model for redeveloping the post-zoning city. Environ Plan B Urban Anal City Sci 49:1005–1023. https://doi.org/10.1177/23998083211041369

Ligmann-Zielinska A, Church RL, Jankowski P (2008) Spatial optimization as a generative technique for sustainable multiobjective land-use allocation. Int J Geogr Inf Sci 22:601–622. https://doi.org/10.1080/13658810701587495

Lindau LA, Hidalgo D, Facchini D (2010) Bus rapid transit in Curitiba, Brazil. Transp Res Rec J Transp Res Board 2193:17–27. https://doi.org/10.3141/2193-03

Lopes MN, Camanho AS (2013) Public green space use and consequences on urban vitality: an assessment of European Cities. Soc Indic Res 113:751–767. https://doi.org/10.1007/s11205-012-0106-9

Lyu G, Bertolini L, Pfeffer K (2020) How does transit-oriented development contribute to station area accessibility? A study in Beijing. Int J Sustain Transp 14:533–543. https://doi.org/10.1080/15568318.2019.1578841

Manasra H, Toledo T (2019) Optimization-based operations control for public transportation service with transfers. Transp Res Part C Emerg Technol 105:456–467. https://doi.org/10.1016/j.trc.2019.06.011

McKenzie G (2020) Urban mobility in the sharing economy: a spatiotemporal comparison of shared mobility services. Comput Environ Urban Syst 79:101418. https://doi.org/10.1016/j.compenvurbsys.2019.101418

Metzinger J (2021) Differences in experiences with the development of mixed-use projects from 2004 and 2017. Front Built Environ 7. https://doi.org/10.3389/fbuil.2021.734149

Mitrany M (2005) High density neighborhoods: who enjoys them? GeoJournal 64:131–140. https://doi.org/10.1007/s10708-005-4099-7

Moreno C, Allam Z, Chabaud D et al (2021) Introducing the "15-Minute City": sustainability, resilience and place identity in future post-pandemic cities. Smart Cities 4:93–111. https://doi.org/10.3390/smartcities4010006

Munn K, Dragićević S (2023) Towards a spatio-temporal multicriteria evaluation method: a suitability analysis of residential units in a 3D urban environment. Trans GIS 27:1830–1845. https://doi.org/10.1111/tgis.13101

Nemeškal J, Ouředníček M, Pospíšilová L (2020) Temporality of urban space: daily rhythms of a typical week day in the Prague metropolitan area. J Maps 16:30–39. https://doi.org/10.1080/17445647.2019.1709577

Neutens T (2015) Accessibility, equity and health care: review and research directions for transport geographers. J Transp Geogr 43:14–27. https://doi.org/10.1016/j.jtrangeo.2014.12.006

Neutens T, Schwanen T, Witlox F, de Maeyer P (2010) Evaluating the temporal organization of public service provision using space-time accessibility analysis. Urban Geogr 31:1039–1064. https://doi.org/10.2747/0272-3638.31.8.1039

Newman G, Lee C, Jourdan D et al (2021) Planning the allocation and reallocation of urban space. J Plan Educ Res 41:258–258. https://doi.org/10.1177/0739456X211022680

Ng CP, Law TH, Wong SV, Kulanthayan S (2017) Relative improvements in road mobility as compared to improvements in road accessibility and economic growth: a cross-country analysis. Transp Policy 60:24–33. https://doi.org/10.1016/j.tranpol.2017.08.004

Niu S, Hu A, Shen Z et al (2019) Study on land use characteristics of rail transit TOD sites in new towns—taking Singapore as an example. J Asian Archit Build Eng 18:16–27. https://doi.org/10.1080/13467581.2019.1586712

Palència L, León-Gómez BB, Bartoll X et al (2020) Study protocol for the evaluation of the health effects of superblocks in Barcelona: The "Salut Als Carrers" (health in the streets) project. Int J Environ Res Public Health 17:2956. https://doi.org/10.3390/ijerph17082956

Pozoukidou G, Angelidou M (2022) Urban Planning in the 15-minute city: revisited under sustainable and smart city developments until 2030. Smart Cities 5:1356–1375. https://doi.org/10.3390/smartcities5040069

Pucher J, Dijkstra L (2003) Promoting Safe walking and cycling to improve public health: lessons from The Netherlands and Germany. Am J Public Health 93:1509–1516. https://doi.org/10.2105/AJPH.93.9.1509

Ramaiah M, Avtar R (2019) Urban green spaces and their need in cities of rapidly urbanizing India: a review. Urban Sci 3:94. https://doi.org/10.3390/urbansci3030094

Ramirez Lopez LJ, Grijalba Castro AI (2020) Sustainability and resilience in smart city planning: a review. Sustainability 13:181. https://doi.org/10.3390/su13010181

Redyantanu BP, Damayanti R (2017) Temporality in a discussion of sustainable architecture. Dimens (J Archit Built Environ) 44:163–170. https://doi.org/10.9744/dimensi.44.2.163-170

Reiner M, McElvaney L (2017) Foundational infrastructure framework for city resilience. Sustain Resilient Infrastruct 2:1–7. https://doi.org/10.1080/23789689.2017.1278994

Ritsema van Eck J, Burghouwt G, Dijst M (2005) Lifestyles, spatial configurations and quality of life in daily travel: an explorative simulation study. J Transp Geogr 13:123–134. https://doi.org/10.1016/j.jtrangeo.2004.04.013

Rode P, da Cruz NF (2018) Governing urban accessibility: moving beyond transport and mobility. Appl Mobilities 3:8–33. https://doi.org/10.1080/23800127.2018.1438149

Rose D, Germain A, Bacqué M et al (2013) 'Social Mix' and neighbourhood revitalization in a transatlantic perspective: comparing local policy discourses and expectations in Paris (France), Bristol (UK) and Montréal (Canada). Int J Urban Reg Res 37:430–450. https://doi.org/10.1111/j.1468-2427.2012.01127.x

Ruiz T, Timmermans H (2006) Changing the timing of activities in resolving scheduling conflicts. Transportation (Amst) 33:429–445. https://doi.org/10.1007/s11116-006-0010-8

Ryckewaert M, Zaman J, De Boeck S (2021) Variable arrangements between residential and productive activities: conceiving mixed-use for urban development in Brussels. Urban Plan 6:334–349. https://doi.org/10.17645/up.v6i3.4274

Salam M (2020) Psychological impacts of architectural design on living spaces. J Art, Archit Built Environ 3. https://doi.org/10.32350/jaabe/2020/32/970

Sánchez Criado T, Cereceda Otárola M (2016) Urban accessibility issues. City 20:619–636. https://doi.org/10.1080/13604813.2016.1194004

Scudellari J, Staricco L, Vitale Brovarone E (2020) Implementing the Supermanzana approach in Barcelona. Critical issues at local and urban level. J Urban des 25:675–696. https://doi.org/10.1080/13574809.2019.1625706

Siu KWM, Wong KSL (2015) Flexible design principles. Facilities 33:588–621. https://doi.org/10.1108/F-02-2014-0021

Smith RJ, Hall T (2013) No time out: mobility, rhythmicity and urban patrol in the twenty-four hour city. Sociol Rev 61:89–108. https://doi.org/10.1111/1467-954X.12055

Southworth M (2016) Learning to make liveable cities. J Urban des 21:570–573. https://doi.org/10.1080/13574809.2016.1220152

Sugiyama T, Carver A, Koohsari MJ, Veitch J (2018) Advantages of public green spaces in enhancing population health. Landsc Urban Plan 178:12–17. https://doi.org/10.1016/j.landurbplan.2018.05.019

Tao H, Wang K, Zhuo L, Li X (2019) Re-examining urban region and inferring regional function based on spatial–temporal interaction. Int J Digit Earth 12:293–310. https://doi.org/10.1080/17538947.2018.1425490

Tian L, Liang Y, Zhang B (2017) Measuring residential and industrial land use mix in the peri-urban areas of China. Land Use Policy 69:427–438. https://doi.org/10.1016/j.landusepol.2017.09.036

Tran M, Draeger C (2021) A data-driven complex network approach for planning sustainable and inclusive urban mobility hubs and services. Environ Plan B Urban Anal City Sci 48:2726–2742. https://doi.org/10.1177/2399808320987093

Trounstine J (2023) You won't be my neighbor: opposition to high density development. Urban Aff Rev 59:294–308. https://doi.org/10.1177/10780874211065776

van den Dobbelsteen A, Broersma S, Fremouw M et al (2019) The Amsterdam energy transition roadmap—introducing the city-zen methodology. Smart Sustain Built Environ 9:307–320. https://doi.org/10.1108/SASBE-05-2019-0065

Vardakas JS, Zenginis I, Zorba N et al (2018) Electrical energy savings through efficient cooperation of urban buildings: the smart community case of superblocks' in Barcelona. IEEE Commun Mag 56:102–109. https://doi.org/10.1109/MCOM.2017.1700542

Vlacic L (2022) The outcome of "The 2021 Shape the Future of ITS" competition [Editor's Column]. IEEE Intell Transp Syst Mag 14:3–4. https://doi.org/10.1109/MITS.2022.3189798

Wan T, Lu W, Sun P (2023) Equity impacts of the built environment in urban rail transit station areas from a transit-oriented development perspective: a systematic review. Environ Res Commun 5:092001. https://doi.org/10.1088/2515-7620/acf8b2

Welch TF (2013) Equity in transport: the distribution of transit access and connectivity among affordable housing units. Transp Policy 30:283–293. https://doi.org/10.1016/j.tranpol.2013.09.020

Wey W-M, Huang J-Y (2018) Urban sustainable transportation planning strategies for livable City's quality of life. Habitat Int 82:9–27. https://doi.org/10.1016/j.habitatint.2018.10.002

Wittowsky D, van der Vlugt A-L, Nadler R (2019) Editorial—accessibility: its role in the sustainable transformation of cities. Appl Mobilities 4:139–141. https://doi.org/10.1080/23800127.2019.1609389

Zhang B, Xie G, Gao J, Yang Y (2014) The cooling effect of urban green spaces as a contribution to energy-saving and emission-reduction: a case study in Beijing, China. Build Environ 76:37–43. https://doi.org/10.1016/j.buildenv.2014.03.003

Zhang Y, Liu L, Wang H (2019) A new perspective on the temporal pattern of human activities in cities: the case of Shanghai. Cities 87:196–204. https://doi.org/10.1016/j.cities.2018.10.002

Zhang S, Zhen F, Kong Y et al (2023) Towards a 15-minute city: a network-based evaluation framework. Environ Plan B Urban Anal City Sci 50:500–514. https://doi.org/10.1177/23998083221118570

Zong H (2023) Discussion on "15-Minute City" from the perspective of neighborhood effect and urban facilities. Lect Notes Educ Psychol Public Media 10:249–255. https://doi.org/10.54254/2753-7048/10/20230224

# 4 Community Engagement

## 4.1 Introduction

Community engagement is a cornerstone of urban planning (Supriyanto et al. 2022), particularly in the development of the **15-minute city**, a model that prioritizes accessibility, proximity, and inclusivity in urban environments. By involving citizens in the urban planning and design process, cities can ensure that developments reflect the unique needs, values, and aspirations of the community. This participatory approach not only enhances the functionality and liveability of urban spaces but also fosters a sense of ownership, accountability, and long-term commitment among residents (Ismail and Said 2015).

In the context of the 15-minute city, meaningful community participation is especially critical. As (Shoina et al. 2024) noted, the success of the 15-minute city hinges on tailoring urban developments to the everyday realities of residents, ensuring that spaces are equitable, responsive, and adaptable to diverse lifestyles and challenges. Active participation helps address key urban challenges, such as accessibility gaps, social inequities, and environmental sustainability, while empowering communities to co-create solutions that improve quality of life for all (Ismail and Said 2015).

By recognizing the critical role of community engagement in shaping urban spaces, this chapter underscores how inclusive and participatory practices can help achieve the ambitious goals of the 15-minute city, fostering urban environments that are not only functional but also enriched by the active involvement of their residents.

This chapter explores the transformative role of community engagement in the urban planning and design, with a focus on the 15-minute city model. It examines how participatory processes ensure that urban spaces are inclusive and responsive, highlights innovative methods for fostering community engagement (e.g. participatory workshops, digital platforms, and collaborative design), and showcases successful community-led initiatives from cities that have embraced this approach. Finally, the chapter provides actionable guidelines for effective participation and discusses how community-based urban design can create more sustainable, resilient, and equitable urban environments.

## 4.2 Community Engagement and Participation

Public and community engagement and participation are essential components of inclusive and sustainable urban planning and design. These practices involve meaningful collaboration between planners and residents to ensure



M. Bencekri and S. Lee, *Sustainable Urban Evolution*, Sustainable Development Goals Series,
https://doi.org/10.1007/978-981-96-5146-7_4

**Fig. 4.1** Tools for community engagement and participation

that urban environments reflect the unique needs, values, and aspirations of local communities. By empowering citizens to actively participate in decision-making, public engagement builds trust, fosters a sense of ownership, and enhances the long-term success of urban initiatives (Geekiyanage et al. 2021).

In the context of the **15-minute city**, community engagement becomes even more significant. This model emphasizes proximity, accessibility, and inclusivity, ensuring that residents can access essential services and amenities within a short walk or bike ride. To achieve this vision, participatory processes permit involving diverse stakeholders and incorporate local knowledge to create urban environments that are equitable, resilient, and adaptable to the rhythms of daily life (Konsti-Laakso and Rantala 2018).

This section explores two critical aspects of community engagement: the **methodologies and tools** used to effectively involve residents in urban planning and design, and the **guidelines for successful participation**, which ensure that participatory processes are impactful, inclusive, and sustainable.

### 4.2.1 Methods and Tools

Effective community engagement requires a diverse set of tools and methods to gather input, foster collaboration, and empower residents to shape their urban environments. These tools not only ensure that a wide range of perspectives is included but also help planners design more equitable and inclusive urban spaces (Fig. 4.1).

#### 4.2.1.1 Public Meetings and Workshops

Public workshops and meetings are essential for creating interactive forums where residents can actively share ideas, provide feedback, and collaborate with planners. These discussions often focus on specific aspects of urban design, such as transportation networks, housing projects, or public spaces.

For example, ***Portland's participatory watershed program*** successfully engaged residents in designing green urban spaces, demonstrating how workshops can address environmental concerns while fostering community ownership (Shandas and Messer 2008).

In the context of the 15-minute city, workshops serve as an essential mechanism for fostering participatory urban planning and ensuring that city design aligns with residents' daily needs. These workshops provide a platform for community members to actively contribute ideas and feedback on key components of the urban environment, including walkability, green spaces, and the strategic placement of essential services. Through structured discussions and collaborative activities, workshops facilitate a bottom-up approach to decision-making that complements top-down planning processes.

#### 4.2.1.2 Focus Groups, Surveys, and Questionnaires

**Focus groups** provide small, diverse groups of residents the opportunity to explore specific urban issues in depth, generating insights into community concerns and innovative solutions (Wibeck et al. 2007). In Los Angeles, for example, focus groups have been used to better understand public perceptions of urban green spaces, leading to more resident-driven urban designs (Gearin and Kahle 2006).

**Surveys and questionnaires** complement focus groups by collecting broad data on community priorities (Jones et al. 2013). Digital tools, such as mobile apps and online forms, make these tools more accessible, enabling large-scale participation. A notable example is Helsinki's digital participatory budgeting platform, which allows residents to directly influence how public funds are allocated, creating more inclusive urban decision-making processes (Tseng et al. 2024).

#### 4.2.1.3 Digital and Emerging Tools

Digital platforms are increasingly central to community engagement, providing accessible, scalable ways for residents to participate in urban planning (Siqueira et al. 2022). Virtual town halls, interactive GIS mapping, and apps for participatory budgeting allow residents to collaborate remotely and offer their input on projects.

For instance, Barcelona's Decidim platform allows residents to debate, propose, and vote on urban initiatives, directly shaping public policies and infrastructure investments (Font-Cot et al. 2023). Similarly, virtual reality simulations are being used in urban design to let residents visualize proposed changes to their neighbourhoods, enabling them to provide informed feedback. These tools are particularly effective in engaging tech-savvy populations and those unable to attend in-person events.

#### 4.2.1.4 Participatory Budgeting

**Participatory budgeting** empowers communities to decide how municipal budgets are allocated, ensuring that public spending reflects local priorities (Wu 2014). In Porto Alegre, Brazil, participatory budgeting has improved governance by creating more transparent, accountable, and responsive decision-making processes (Calisto Friant 2019).

**Co-design workshops** involve residents directly in the design of urban spaces, fostering collaboration between planners, architects, and community members (Ng et al. 2024). These workshops are particularly useful for designing public spaces, such as parks and plazas, where local knowledge can inform design decisions that enhance usability and accessibility.

### 4.2.2 Effective Public Participation

For community engagement to be effective, participatory processes must follow key principles: transparency, accountability, awareness, and inclusivity (Fig. 4.2). These principles ensure that public participation is meaningful, equitable, and capable of addressing the diverse needs of urban residents (Rohman Kusmayad et al. 2024; Damanik et al. 2024; Noya 2024).

#### 4.2.2.1 Transparency

Transparency builds trust by ensuring that all aspects of the planning process, goals, timelines, and decision-making criteria, are communicated openly. Open communication fosters long-term public support for urban initiatives (Konsti-Laakso and Rantala 2018).

For instance, Freiburg's sustainable urban development succeeded because project objectives and progress were clearly communicated to the public, encouraging sustained community involvement (Mössner 2016).

**Fig. 4.2** Effective community engagement and participation

#### 4.2.2.2 Accountability

Accountability is equally vital as it ensures that planners are answerable to the public and committed to acting on the input they receive (Tasan-Kok et al. 2019). When community members see that their feedback is not only considered but also acted upon, their trust in the planning process deepens (Jukur et al. 2023). A study on participatory budgeting in Porto Alegre demonstrated that transparent and accountable governance improved relations between citizens and the government, reducing local corruption and clientelism (Calisto Friant 2019). This has led to more responsive urban policies that reflect the community's needs.

#### 4.2.2.3 Awareness

Awareness is another fundamental component, as meaningful participation requires that residents have the necessary knowledge and resources to engage effectively (Laurian 2003). For instance, Denmark's participatory design "FabLabSchool," which aims at integrating digital technologies into primary and secondary education, focuses on ensuring that participants understand their roles and the potential impact of their contributions (Smith and Iversen 2018).

#### 4.2.2.4 Inclusivity

Inclusivity ensures that all segments of the population, especially marginalized and underrepresented groups, can participate in the planning process (van den Berg et al. 2020). Seattle's inclusive planning, which specifically targeted marginalized communities, resulted in more equitable outcomes and a better alignment with community needs (Sirianni 2007).

Community engagement and public participation are indispensable for creating equitable, inclusive, and sustainable urban environments. By leveraging diverse methodologies, such as digital tools, participatory budgeting, and co-design workshops, and adhering to principles of transparency, accountability, awareness, and inclusivity, planners can ensure that residents play an active role in shaping their urban futures.

## 4.3 Participatory Design in a 15-Minute City

Participatory design plays a central role in realizing the vision of the **15-minute city**, a model that seeks to create accessible, inclusive, and sustainable urban environments. This

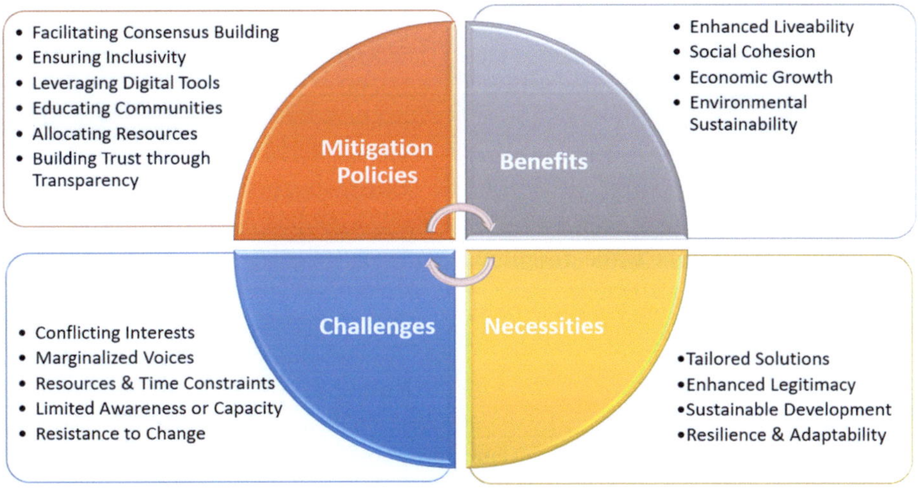


**Fig. 4.3** Guideline for a participatory design of a 15-minute city

approach enables planners to harness local knowledge, tailor solutions to the unique characteristics of neighbourhoods, and address the diverse challenges faced by urban communities. Participatory processes not only enhance the legitimacy and adaptability of urban projects but also contribute to long-term sustainability and resilience by aligning initiatives with community priorities.

This section explores the necessity, benefits, and challenges of participatory design in the 15-minute city context, while also providing recommendations for overcoming obstacles to maximize its potential (Fig. 4.3).

### 4.3.1 Necessity

Participatory design is a critical component of the **15-minute city** concept, ensuring that urban spaces are designed to meet the specific needs of residents while fostering inclusivity and sustainability. Unlike top-down planning approaches, participatory design incorporates local knowledge, values, and lived experiences into urban planning, leading to more responsive and equitable outcomes.

- **Tailored Solutions**: Community input ensures that urban planning initiatives address the unique needs of different neighbourhoods. Planners gain insights into localized challenges, such as accessibility barriers or infrastructure gaps, enabling them to contextually craft relevant solutions (Brookfield 2017). For example, in Paris, community involvement has been integral to shaping the city's 15-minute neighbourhoods, ensuring that services align with residents' daily routines (Kohli 2024).
- **Enhanced Legitimacy**: Involving residents in the planning process strengthens transparency and accountability, building trust between communities and urban authorities (Lekomtseva et al. 2020). Residents are more likely to support and maintain urban projects when they feel a sense of ownership and see their input reflected in the outcomes.
- **Sustainable Development**: Participatory design aligns urban initiatives with local environmental, social, and economic priorities, fostering long-term sustainability (Mia et al. 2022). Community-driven projects often prioritize green infrastructure, renewable energy, and local economic growth, resulting in more sustainable urban environments (Kelvin 2024).
- **Resilience and Adaptability**: Collaborative planning enhances the city's ability to adapt to evolving challenges (Nassar et al. 2025). For instance, in New York City, community

involvement in disaster preparedness initiatives strengthened resilience and improved the city's emergency response capabilities (Rivera et al. 2019).

In summary, participatory design is essential for tailoring urban planning initiatives to the specific needs of residents while fostering legitimacy, sustainability, and resilience.

### 4.3.2 Benefits

A community-based approach to urban design within the 15-minute city framework delivers transformative benefits across social, economic, and environmental dimensions.

- **Enhanced Liveability**: Participatory processes ensure that urban spaces reflect residents' needs and aspirations, improving daily living conditions (Kohli 2024). Co-created designs result in better public amenities, walkable neighbourhoods, and improved access to essential services.
- **Social Cohesion**: Collaborative planning fosters trust, belonging, and mutual understanding among residents and foster social cohesion (Butu et al. 2023). For example, in Barcelona, community-driven initiatives have turned public spaces into vibrant hubs of interaction, strengthening neighbourhood bonds and creating a sense of community (Eizaguirre Anglada 2022).
- **Economic Growth**: Locally informed designs support small businesses, attract visitors, and stimulate economic activity (Matiku et al. 2020). For instance, the Makuleke Contractual Park Community Project in South Africa demonstrates how participatory design can drive sustainable economic development by involving communities in tourism and conservation (Mohapatra 2023).
- **Environmental Sustainability**: Community participation ensures the prioritization of environmental sustainability (Markelj et al. 2024). Engaged communities are more likely to support and maintain environmental planning initiatives, such as green spaces, renewable energy, and sustainable infrastructure (Tuanaya 2024). In Copenhagen, community participation has been instrumental in creating bike-friendly streets and extensive parks, significantly reducing the city's carbon footprint (Treija et al. 2023). Such initiatives illustrate how participatory design can lead to greener, healthier urban environments.

Participatory design empowers residents to shape their surroundings while fostering social, economic, and environmental well-being.

### 4.3.3 Challenges

Despite its many advantages, participatory design faces several challenges that can undermine its effectiveness if not properly addressed:

- **Diverse and Conflicting Interests**: Communities are often diverse, with residents holding different, and sometimes conflicting, priorities. For example, while some residents may prioritize housing, others may advocate for green spaces or cultural amenities. Balancing these competing demands is a significant challenge for planners (Konsti-Laakso and Rantala 2018).
- **Marginalized Voices**: Vulnerable groups, such as low-income residents, individuals with disabilities, or minority communities, are often underrepresented in participatory processes due to barriers such as language, digital literacy, or time constraints (van den (Berg et al. 2020)).
- **Resources and Time Constraints**: Participatory design can be resource-intensive, requiring significant time, funding, and staff (Kraff and Jernsand 2022). This poses challenges for cities with tight budgets or fast-tracked urban projects.
- **Limited Awareness or Capacity**: Many residents lack the technical knowledge or confidence to meaningfully engage in urban planning discussions (Geekiyanage et al. 2020).

Without proper education and support, participation can be superficial, leading to ineffective or inequitable outcomes (Laurian 2003).

- **Resistance to Change**: Some community members may resist participatory processes, particularly if they distrust local governments or feel disengaged from decision-making.

### 4.3.4 Mitigation Policies

To overcome these challenges and maximize the benefits of participatory design, urban planners can adopt the following strategies:

- **Facilitating Consensus Building**: Use structured methods, such as deliberative democracy or scenario planning workshops, to address conflicting interests. Facilitators can guide discussions to identify shared priorities and reach equitable solutions.
- **Ensuring Inclusivity**: Actively engage marginalized groups through targeted outreach, multilingual materials, flexible participation options (e.g. in-person and online formats), and accessible venues. For example, **Seattle's inclusive planning efforts** have successfully engaged underrepresented populations, leading to more equitable urban outcomes (Sirianni 2007).
- **Leveraging Digital Tools**: Use participatory platforms, such as **GIS mapping**, digital surveys, and virtual co-design workshops, to streamline engagement processes. Tools like **Barcelona's Decidim platform** enable broader participation while reducing logistical barriers.
- **Educating Communities**: Provide educational resources, training sessions, and workshops to build residents' capacity to engage effectively in urban planning. Denmark's **FabLabSchool** is a model for empowering communities through education (Smith and Iversen 2018).
- **Allocating Resources**: Secure dedicated funding and staff to support participatory initiatives, ensuring that community engagement processes are well-organized and impactful.
- **Building Trust Through Transparency**: Regularly update residents on project progress, demonstrate how their input has influenced decisions, and ensure accountability in decision-making. Transparency fosters trust and long-term community engagement (Mössner 2016).

Participatory design is essential for realizing the vision of the **15-minute city**, ensuring that urban planning initiatives are inclusive, equitable, and reflective of residents' needs. By actively involving communities, planners can create tailored solutions that enhance liveability, social cohesion, economic growth, and environmental sustainability.

However, to fully realize the potential of participatory design, planners must address challenges such as conflicting interests, marginalized voices, and resource constraints. Through strategies like inclusivity, education, digital tools, and transparency, cities can mitigate these obstacles and empower residents to shape their urban environments. Ultimately, participatory design fosters a sense of shared ownership, creating vibrant, resilient, and people-centred cities that align with the principles of the 15-minute city.

## 4.4 Conclusion

Community engagement and participatory design lie at the heart of building equitable, sustainable, and resilient urban environments, particularly within the framework of the ***15-minute city***. By actively involving residents in urban planning and design, cities can create spaces that are not only functional but also deeply reflective of the needs, aspirations, and lived experiences of their communities. This approach fosters trust, transparency, and a shared sense of ownership, ensuring long-term support and success for urban initiatives (Fig. 4.2).

A wide range of methodologies and tools (Fig. 4.1) empowers residents to participate meaningfully in shaping their neighbourhoods. Public workshops, focus groups, surveys, and digital platforms ensure that diverse voices are included, enabling planners to capture the perspectives of both the majority and marginalized groups. For example, Helsinki's participatory budgeting platform and Barcelona's Decidim platform have successfully broadened engagement through accessible, tech-enabled participation. Such methods ensure inclusivity and encourage the co-creation of urban solutions that align with the principles of the 15-minute city, where essential services and amenities are accessible within short walking or cycling distances.

Beyond participation tools, community-led initiatives further empower residents to directly shape their urban environments. These initiatives, ranging from small-scale projects like community gardens to larger efforts like neighbourhood revitalization programs, foster stronger social cohesion, economic vitality, and environmental sustainability (Fig. 4.3). For instance, Portland's Greenways program created safer, more connected streets for pedestrians and cyclists, while Barcelona's Superblocks transformed car-dominated streets into vibrant pedestrian-friendly zones. Such initiatives highlight the transformative potential of community-driven planning in addressing urban challenges, from reducing car dependency to enhancing public spaces.

However, participatory design is not without its challenges (Fig. 4.3). Diverse and sometimes conflicting interests among residents can complicate decision-making, while marginalized voices often face barriers to engagement due to language, digital divides, or lack of representation. Additionally, resource and time constraints may limit the scope of participatory efforts, particularly in fast-paced urban projects. Planners must also contend with resistance to change, particularly in communities that distrust government processes or fear disruptions to established norms.

To overcome these obstacles, cities must adopt mitigation strategies which ensure that participation is inclusive, impactful, and equitable (Fig. 4.3). Targeted outreach, such as multilingual resources and flexible participation formats, can engage underrepresented groups, while digital tools like participatory GIS mapping and virtual co-design platforms can broaden accessibility. Educational campaigns and capacity-building workshops can empower residents to engage meaningfully in technical discussions. Regular feedback loops, where planners demonstrate how community input has shaped outcomes, are crucial for building trust and fostering long-term commitment.

Ultimately, participatory design offers a ***holistic framework*** for addressing the complexities of urban life. By embedding inclusivity, transparency, and adaptability into the planning process, cities can create environments that are resilient to future challenges, from climate change to social inequities.

Participatory design is a transformative planning approach which ensures that the 15-minute city becomes a reality. By integrating community inputs at every stage, cities can align development with residents' needs, fostering a shared sense of responsibility for urban growth. This participatory model is essential for building equitable, sustainable, and resilient cities that prioritize quality of life for all residents, today and in the future.

## References

Brookfield K (2017) Getting involved in plan-making: Participation in neighbourhood planning in England. Environ Plan C Polit Sp 35:397–416. https://doi.org/10.1177/0263774X16664518

Butu HM, Hassan MM, Hashim AH, Ahmad N (2023) Perceptions of insecurity and social cohesion in building resilience among residents in North-East Nigeria. Int J Acad Res Bus Soc Sci 13. https://doi.org/10.6007/IJARBSS/v13-i12/20409

Calisto Friant M (2019) Deliberating for sustainability: lessons from the Porto Alegre experiment with participatory budgeting. Int J Urban Sustain Dev 11:81–99. https://doi.org/10.1080/19463138.2019.1570219

De Siqueira G, Malaj S, Hamdani M (2022) Digitalization, participation and interaction: towards more inclusive tools in urban design—A literature review. Sustainability 14:4514. https://doi.org/10.3390/su14084514

Damanik FHS, Rerung A, Lubis MDA, et al (2024) Community participation in the public decision process: realizing better governance in public administration. Glob Int J Innov Res 1:88–96. https://doi.org/10.59613/global.v1i2.12

Eizaguirre Anglada S (2022) Cultural commons as a key for bottom-linked policies. An exploration of the support for public and community partnerships in Barcelona. w@terfront Public ArtUrban Des Particip Regen 64:3–39. https://doi.org/10.1344/waterfront2021.63.12.01

Font-Cot F, Lara-Navarra P, Serradell-Lopez E (2023) Digital transformation policies to develop an effective startup ecosystem: the case of Barcelona. Transform Gov People Process Policy 17:344–355. https://doi.org/10.1108/TG-01-2023-0006

Gearin E, Kahle C (2006) Teen and adult perceptions of urban green space Los Angeles. Child Youth Environ 16:25–48. https://doi.org/10.1353/cye.2006.0037

Geekiyanage D, Fernando T, Keraminiyage K (2020) Assessing the state of the art in community engagement for participatory decision-making in disaster risk-sensitive urban development. Int J Disaster Risk Reduct 51:101847. https://doi.org/10.1016/j.ijdrr.2020.101847

Geekiyanage D, Fernando T, Keraminiyage K (2021) Mapping participatory methods in the urban development process: a systematic review and case-based evidence analysis. Sustainability 13:8992. https://doi.org/10.3390/su13168992

Ismail WAW, Said I (2015) Integrating the community in urban design and planning of public spaces: a review in malaysian cities. Procedia Soc Behav Sci 168:357–364. https://doi.org/10.1016/j.sbspro.2014.10.241

Jones T, Baxter M, Khanduja V (2013) A quick guide to survey research. Ann R Coll Surg Engl 95:5–7. https://doi.org/10.1308/003588413X13511609956372

Jukur S, Wiltgen Georgi N, Tolhurst R et al (2023) Catalysing effective social accountability systems through community participation. Cities Heal 7:982–990. https://doi.org/10.1080/23748834.2023.2240478

Kelvin KHW (2024) Enhancing sustainable development through community engagement. Int J Multidiscip Res 6. https://doi.org/10.36948/ijfmr.2024.v06i04.25621

Konsti-Laakso S, Rantala T (2018) Managing community engagement: a process model for urban planning. Eur J Oper Res 268:1040–1049. https://doi.org/10.1016/j.ejor.2017.12.002

Kohli A (2024) Community engagement and participation in planning of a city. Int J Multidiscip Res 6. https://doi.org/10.36948/ijfmr.2024.v06i03.23793

Kraff H, Jernsand EM (2022) Time as an issue of power in participatory design

Laurian L (2003) A prerequisite for participation. J Plan Educ Res 22:257–269. https://doi.org/10.1177/0739456X02250316

Lekomtseva AA, Khatskelevich AN, Gimranova GA (2020) Place-making: application in the city planning process. Ekon I Upr Probl RESHENIYA 1:47–54. https://doi.org/10.36871/ek.up.p.r.2020.05.01.007

Markelj J, Petrovski A, Harmathy N, Kazak J (2024) Participatory design for sustainability, pp 60–67

Matiku S, Zuwarimwe J, Tshipala N (2020) Community-driven tourism projects' economic contribution to community livelihoods—A case of Makuleke contractual park community tourism project. Sustainability 12:8230. https://doi.org/10.3390/su12198230

Mia MT, Islam M, Sakin J, Al-Hamadi J (2022) The role of community participation and community-based planning in sustainable community development. Asian People J 5:31–41. https://doi.org/10.37231/apj.2022.5.1.296

Mohapatra B (2023) Community participation in public space planning and management: cases of Indian cities. Interdisciplinary perspectives on sustainable development. CRC Press, London, pp 136–141

Mössner S (2016) Sustainable urban development as consensual practice: post-politics in Freiburg, Germany. Reg Stud 50:971–982. https://doi.org/10.1080/00343404.2015.1102875

Nassar S, Hossain MB, Naár ZT, Vasa L (2025) Factors affecting humanitarian resilience: a mediating approach of community engagement. Decis Mak Appl Manag Eng 8:108–131. https://doi.org/10.31181/dmame8120251106

Ng P, Zhu S, Li Y, van Ameijde J (2024) Digitally gamified co-creation: enhancing community engagement in urban design through a participant-centric framework. Des Sci 10:e17. https://doi.org/10.1017/dsj.2024.17

Noya J (2024) Community participation makes equal welfare. J Soc Interact Humanit 3:211–222. https://doi.org/10.55927/jsih.v3i2.10253

Rivera L, Pagaoa M, Morgenthau BM et al (2019) Participation in community preparedness programs in human services organizations and faith-based organizations—New York City, 2018. MMWR Morb Mortal Wkly Rep 68:757–761. https://doi.org/10.15585/mmwr.mm6835a2

Rohman Kusmayad RC, Wen IM, Jatmikowati SH (2024) Community participation in village development. Int J Res Soc Sci Humanit 05:124–145. https://doi.org/10.47505/IJRSS.2024.6.10

Supriyanto E, Saputra J, Rachmawati M, Nugroho F (2022) Community participation-based smart city development. In: Proceedings of the first multidiscipline international conference, MIC 2021, 30 Oct 2021, Jakarta, Indonesia. EAI

Treija S, Stauskis G, Korolova A, Bratuskins U (2023) Community engagement in urban experiments: joint effort for sustainable urban transformation. Landsc Archit art 22:89–97. https://doi.org/10.22616/j.landarchart.2023.22.09

Tuanaya W (2024) Building eco-friendly cities: government-community collaboration in shaping sustainable urban waste management. Baileo J Sos Hum 1:281–292. https://doi.org/10.30598/baileofisipvol1iss3pp281-292

Shandas V, Messer WB (2008) Fostering green communities through civic engagement: community-based environmental stewardship in the Portland area. J Am Plan Assoc 74:408–418. https://doi.org/10.1080/01944360802291265

Shoina M, Voukkali I, Anagnostopoulos A et al (2024) The 15-minute city concept: the case study within a neighbourhood of Thessaloniki. Waste Manag Res J Sustain Circ Econ 42:694–710. https://doi.org/10.1177/0734242X241259926

Sirianni C (2007) Neighborhood planning as collaborative democratic design. J Am Plan Assoc 73:373–387. https://doi.org/10.1080/01944360708978519

Smith RC, Iversen OS (2018) Participatory design for sustainable social change. Des Stud 59:9–36. https://doi.org/10.1016/j.destud.2018.05.005

Tasan-Kok T, van den Hurk M, Özogul S, Bittencourt S (2019) Changing public accountability mechanisms in the governance of Dutch urban regeneration. Eur Plan Stud 27:1107–1128. https://doi.org/10.1080/09654313.2019.1598017

Tseng Y-S, Becker C, Roikonen I (2024) Dialectical approach to unpacking knowledge-making for digital urban democracy: a critical case of Helsinki-based e-participatory budgeting. Urban Stud 61:112–129. https://doi.org/10.1177/00420980231175247

van den Berg AC, Giest SN, Groeneveld SM, Kraaij W (2020) Inclusivity in online platforms: recruitment strategies for improving participation of diverse sociodemographic groups. Public Adm Rev 80:989–1000. https://doi.org/10.1111/puar.13215

Wibeck V, Dahlgren MA, Öberg G (2007) Learning in focus groups. Qual Res 7:249–267. https://doi.org/10.1177/1468794107076023

Wu Y (2014) Participatory budgeting. In: The Changing policy-making process in Greater China. Routledge, pp 59–77

# Smart 15-Minute Cities

# 5

## 5.1 Introduction

The ***15-minute city*** is a transformative urban model that prioritizes accessibility, equity, and sustainability by ensuring that essential services, such as health care, education, work, and recreation, are within a 15-minute walk or bike ride of every resident's home (Allam et al. 2022a). By promoting localized, community-focused development, this approach reduces car dependency, minimizes environmental impact, and fosters vibrant, human-centred neighbourhoods (Moreno et al. 2021; Allam et al. 2022b). However, the integration of ***smart city technologies*** elevates this model further, enabling cities to become more efficient, inclusive, and adaptable to the challenges of modern urban life (Allam et al. 2022c).

Smart technologies, including information and communication technologies (ICT), Internet of Things (IoT) devices, and real-time data analytics, allow cities to optimize public services, transportation networks, and resource management. These advancements transform cities into dynamic ecosystems capable of responding to residents' needs in real time (Bifulco et al. 2016; Yeh 2017). For instance, smart mobility solutions can streamline transportation, reduce congestion, and enhance connectivity, allowing for seamless movement within and between different neighbourhoods (Oladimeji et al. 2023). Similarly, data-driven resource management systems optimize the allocation of energy, water, and waste services, significantly reducing environmental footprints and promoting long-term sustainability (Wu et al. 2022).

In addition to enhancing efficiency, smart technologies play a pivotal role in fostering social inclusivity and equity. Digital platforms empower residents to participate in decision-making processes, voice their needs, and contribute to the urban planning process (Sha et al. 2024). For example, platforms like Barcelona's Decidim allow citizens to propose, debate, and vote on urban initiatives (Borge et al. 2023), ensuring that the voices of diverse populations are heard and reflected in public policy (Van Aeken 2016). This participatory approach ensures that smart 15-minute cities are not only technologically advanced but also equitable and responsive to the needs of all residents, including marginalized groups.

However, implementing smart 15-minute cities is not without its challenges. ***Privacy concerns***, ***digital divides***, and the ***high costs of infrastructure and maintenance*** must be carefully managed to ensure that these advancements benefit all residents rather than exacerbating existing inequities (Vassilakopoulou and Hustad 2023). For instance, urban planners must address the risk of excluding low-income or digitally underserved populations by ensuring equitable access to smart tools and services. Furthermore, robust data governance and



M. Bencekri and S. Lee, *Sustainable Urban Evolution*, Sustainable Development Goals Series,
https://doi.org/10.1007/978-981-96-5146-7_5

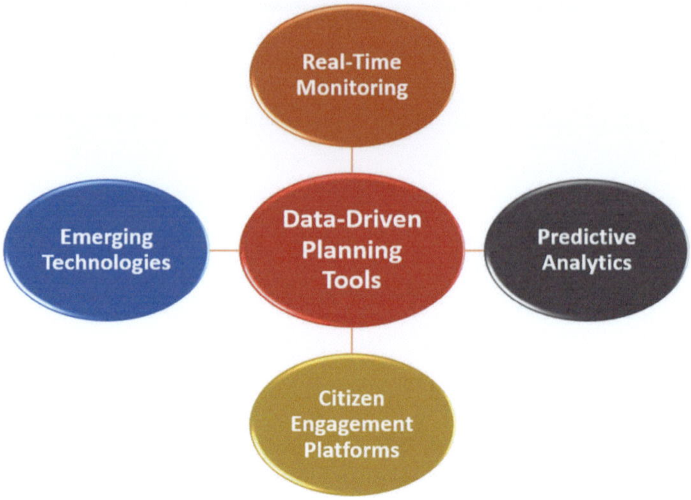


**Fig. 5.1** Data-driven planning tools

privacy policies are essential to building public trust in these technologies.

This chapter examines the intersection of localized urban design and smart technologies within the 15-minute city framework. It explores how digital innovations, such as smart mobility, resource management, and community engagement platforms, can amplify the impact of the 15-minute city. At the same time, it highlights the challenges of integrating these technologies and offers strategies to mitigate them. Ultimately, the Smart 15-Minute City represents a forward-looking approach to urban development, offering a blueprint for creating resilient, sustainable, and inclusive cities that prioritize human well-being while leveraging the power of technology.

## 5.2 Data-Driven City Planning

Data-driven city planning is revolutionizing how urban environments are managed and developed, leveraging advanced technologies to enhance decision-making, resource allocation, and service delivery (Wu et al. 2022). By integrating **real-time monitoring**, **predictive analytics**, **citizen engagement platforms**, and **emerging technologies** (Fig. 5.1), this approach enables cities to become smarter, more sustainable, and resilient.

Data-driven planning is not merely about collecting data, it is about transforming data into actionable insights that optimize urban systems, empower communities, and address the challenges of urbanization, climate change, and social inequities. However, while the potential benefits are immense, challenges such as privacy concerns, digital divides, governance, and implementation costs must be carefully managed to ensure equitable and inclusive outcomes.

### 5.2.1 Real-Time Monitoring and Management

**Real-time monitoring** is the backbone of data-driven urban planning, utilizing sensors, IoT devices, and connected systems to continuously track and optimize urban services and infrastructure (Lv et al. 2020). This technology enables cities to identify and respond to issues such as traffic congestion, waste management inefficiencies, and energy overuse in real time, ensuring smoother and more efficient urban operations.

For example, Barcelona's IoT network manages diverse systems, including street lighting, waste collection, and traffic flow, resulting in substantial operational cost savings, improved sustainability, and enhanced urban functionality (Kamel Boulos and Al-Shorbaji

2014). Similarly, smart traffic management systems in Singapore adjust traffic signals dynamically based on congestion data, reducing travel times and lowering vehicle emissions (Sultana et al. 2024).

Real-time monitoring also plays a critical role in public safety. For instance, smart surveillance systems powered by IoT and AI can detect emergencies such as accidents or natural disasters, enabling rapid response and minimizing damage (Bano et al. 2024). By continuously monitoring urban environments, cities can become more responsive, adaptable, and capable of meeting residents' needs in real time.

### 5.2.2 Predictive Analytics

While real-time monitoring focuses on immediate responsiveness, **predictive analytics** empowers cities to anticipate and plan for future challenges (Wu et al. 2022). By analysing vast amounts of historical and real-time data, urban planners can forecast traffic patterns, energy demands, population growth, and climate-related risks with precision (Cesario 2023).

For instance, predictive models allow cities to allocate resources more effectively by identifying areas that will require infrastructure upgrades or renewable energy installations. In Amsterdam, predictive analytics is used to forecast energy usage in residential neighbourhoods, enabling better distribution of renewable solar energy through smart grids (Wisdom Samuel (Wisdom Samuel Udo et al. 2023). Additionally, predictive models help cities plan for extreme weather events by identifying vulnerable areas and implementing adaptive strategies, such as flood-resilient drainage systems or heat-resistant urban design.

This shift from reactive to proactive planning enables cities to optimize resource allocation, reduce environmental impacts, and enhance the overall resilience of urban systems (Hadiyana and Ji-hoon 2024).

### 5.2.3 Citizen Engagement Digital Platforms

To ensure inclusivity in data-driven planning, **citizen engagement platforms** provide residents with the tools to actively shape their urban environments. These platforms allow residents to report issues, provide feedback, and participate in decision-making, ensuring that different opinions and perspectives are represented in urban planning (Bastos et al. 2022).

One notable example is the UK's FixMyStreet platform enables residents to report local issues like potholes or broken infrastructure, ensuring timely resolutions and strengthening trust in municipal governance (King and Brown 2007).

By integrating citizen-generated data into urban systems, cities can bridge the gap between residents and decision-makers, making urban planning more participatory and responsive. However, to maximize the potential of these platforms, cities must address challenges like digital literacy gaps, language barriers, and limited internet access to ensure that all residents, especially marginalized groups, can participate equally.

### 5.2.4 Emerging Technologies

Emerging technologies such as **AI**, **machine learning**, **digital twins**, and **blockchain** are further advancing data-driven city planning.

- **AI and machine learning** enhance predictive analytics by identifying hidden patterns in complex datasets, enabling cities to make smarter, data-informed decisions (Chai et al. 2024).
- **Digital twins** create virtual models of cities, allowing planners to simulate infrastructure changes or predict the impacts of urban interventions before implementation (Adreani et al. 2024).

- **Blockchain technology** improves transparency and security in data sharing, particularly in citizen engagement platforms and participatory budgeting processes (Sousa 2023).

## 5.3 Smart Sustainability, Resilience, and Equity

Smart city planning is rooted in the principles of **sustainability**, **resilience**, and **equity**, creating urban environments that are efficient, inclusive, and adaptable to future challenges (Ramirez Lopez and Grijalba Castro 2020). These principles address the complex demands of urbanization, including climate change, social inequality, and rapid population growth, while leveraging technology to create liveable, human-centred cities. The success of smart city planning lies in integrating these principles holistically, ensuring that they work together to achieve long-term urban well-being.

### 5.3.1 Smart Environmental Sustainability

**Environmental sustainability** is a cornerstone of smart city planning, focusing on minimizing ecological footprints while enhancing resource efficiency (Laffta and Al-rawi 2018). By adopting **green technologies**, **circular economy models**, and **nature-based solutions**, cities can mitigate climate impacts, conserve resources, and create healthier environments.

Some key strategies include (Laffta and Al-rawi 2018; Nižetić et al. 2019):

- Renewable Energy Systems: Transitioning to solar, wind, and geothermal energy, supported by smart grids and energy storage systems, reduces dependency on fossil fuels (Kalair et al. 2021). For example, Amsterdam's smart grid enables energy sharing among residents, creating a decentralized and efficient system (Pandey et al. 2024).
- Sustainable Transportation: Electric vehicles (EVs), autonomous public transit, and active mobility options like cycling and walking help reduce urban emissions (Wang 2024). Berlin, for instance, has invested in pressure-sensitive floors in cycle paths to produce electricity through friction from bicycle tyres, promoting cycling infrastructure sustainability (Gorges et al. 2020)
- Circular Economy Models: Emphasizing waste reduction, recycling, and resource reuse minimizes environmental degradation (Sanguino et al. 2020). San Francisco's zero waste program, which combines recycling and composting, exemplifies how cities can create closed-loop systems (Ai 2024).
- Nature-Based Solutions and Urban Biodiversity: Integrating green infrastructure, such as parks, wetlands, and green roofs, enhances urban resilience while fostering biodiversity (Filazzola et al. 2019). Singapore's Gardens by the Bay demonstrates how urban greening can mitigate heat islands, improve air quality, and create recreational spaces (Wu et al. 2024).

Additionally, sustainable urban planning improves public health by reducing pollution-related illnesses, encouraging physical activity, and providing mental health benefits through access to green spaces. By prioritizing sustainability, smart cities can create environments that are both ecologically responsible and conducive to long-term well-being.

### 5.3.2 Smart Urban Resilience

**Resilience** ensures that cities can withstand, adapt to, and recover from disruptions, including natural disasters, economic shocks, and public health crises (Mottahedi et al. 2021). **Resilient cities** leverage **smart technologies** and **adaptive policies** to protect physical infrastructure and support social stability.

Core strategies for resilience include:

- **Climate-Resilient Infrastructure**: Cities must invest in adaptive advanced infrastructure, such as flood-proof buildings, permeable pavements, and green drainage systems, to combat extreme weather events. For example, Tokyo's flood tunnels protect the city from typhoon-related damage (Nakamura and Oosawa 2021).
- **Smart Disaster Management**: Real-time monitoring and AI-driven systems enable early detection and rapid response to emergencies. Mexico City's seismic alert system provides real-time warnings for earthquakes, saving lives and minimizing disruptions (Suárez 2022).
- **Energy Resilience**: Distributed energy systems, such as microgrids, ensure that critical facilities (e.g. hospitals, transportation hubs) remain operational during power outages (Qiu et al. 2024).
- **Community Resilience**: Building social cohesion through community-based disaster preparedness programs ensures that vulnerable populations are not left behind (Gil-Rivas and Kilmer (Gil-Rivas and Kilmer 2016).

However, resilience planning must be equitable. (Fainstein 2018) highlights how New York's resilience strategies, such as coastline fortifications, disproportionately benefited wealthy neighbourhoods while leaving marginalized communities at risk. To address this, cities must prioritize vulnerable populations, incorporating equitable disaster relief funding, affordable retrofitting programs, and early-warning systems tailored to underserved communities.

### 5.3.3 Smart Social Equity

**Social equity** ensures that technological advancements and urban innovations benefit all residents, regardless of socioeconomic status, ability, or demographic background (Lee et al. 2023). This principle addresses systemic disparities and promotes inclusivity in urban planning.

Key strategies include:

- **Bridging the Digital Divide**: Expanding affordable internet access and launching digital literacy programs ensure that all residents can benefit from smart city initiatives. For instance, India's Digital India initiative has provided internet access to rural populations, reducing technological disparities (Patel 2024).
- **Inclusive Governance**: Participatory platforms, such as Helsinki's participatory budgeting system, allow marginalized groups to influence urban planning decisions, fostering transparency and inclusivity (Ertiö et al. 2019).
- **Equitable Resource Allocation**: Targeted investments in underserved neighbourhoods address disparities in housing, transportation, and public services. Portland's smart mobility initiative, for example, prioritizes equity by directing resources to low-income communities (Golub et al. 2019).
- **Universal Design Principles**: Designing accessible urban spaces, such as step-free public transit, tactile paving, and inclusive public facilities, ensures that people with disabilities, children, and the elderly can fully participate in urban life (Marchigiani 2023).

Achieving equity in smart cities creates **social cohesion**, reduces inequality, and ensures that all residents can thrive in a technologically advanced urban environment.

### 5.3.4 Smart Integrated Model

The integration of **sustainability**, **resilience**, and **equity** into a unified framework is critical for smart city success, as these principles are deeply interconnected (Narain 2024; Ammar et al. 2024).

Resilience depends on sustainability because climate-resilient infrastructure relies on green technologies to withstand environmental challenges (Roostaie et al. 2019). Similarly, equity enhances resilience by ensuring inclusive disaster preparedness that protects vulnerable populations (Rendon et al. 2021), while sustainability promotes equity by improving access to clean air, water, and public spaces for everyone through green infrastructure (Leach et al. 2018).

Medellín city exemplifies how these principles can converge. Its cable car system not only reduces emissions but also connects marginalized communities to economic opportunities (Brand and Dávila (Brand and Dávila 2011), while the city's green spaces and social programs enhance environmental quality and foster inclusivity (Brand and Dávila 2011). This transformation demonstrates that by integrating sustainability, resilience, and equity, cities can create vibrant, adaptive, and inclusive environments, offering a model for urban development worldwide.

### 5.3.5 Challenges and Recommendations

#### 5.3.5.1 Financial Constraints

Budget limitations are a major barrier to smart city initiatives (Yu 2024). To address this, cities can form public–private partnerships (PPPs) that leverage private sector investment in exchange for shared benefits, such as co-managing urban infrastructure (Radchenko 2022).

Additionally, securing international grants from organizations like the World Bank or climate funds can provide vital financial support, particularly for sustainability-focused projects (Ahamer 2021). Clear frameworks for managing PPPs can ensure accountability and equitable resource allocation.

#### 5.3.5.2 Digital Inequities

The digital divide prevents equitable participation in smart city initiatives (Kolotouchkina et al. 2024). To tackle this, cities must prioritize expanding affordable internet access, such as through community Wi-Fi networks or subsidies for low-income households.

In parallel, digital literacy programs should be developed, targeting marginalized groups like seniors or low-income communities, to ensure that they can fully engage with digital platforms (Choudhary and Bansal 2022). For instance, cities could collaborate with educational institutions or NGOs to deliver workshops and training programs tailored to specific populations.

#### 5.3.5.3 Data Privacy Concerns

As smart city projects rely on vast amounts of data, privacy concerns can erode public trust (Braun et al. 2018). Implementing robust governance frameworks is essential, including clear regulations for data collection, use, and storage. Cities should adopt privacy-by-design principles, use anonymization techniques, and establish independent oversight bodies to monitor data practices. Public transparency about how resident data is used, combined with opt-in models, can further reassure citizens and encourage participation (Matheus and Janssen (Matheus and Janssen 2020; Cerrillo-Martínez and Casadesús-de-Mingo 2021).

#### 5.3.5.4 Resistance to Change

Resistance to new technologies or policies is often rooted in a lack of understanding or mistrust (Matheus and Janssen (Matheus and Janssen 2020; DuBose and Mayo 2020). Early and transparent community engagement is critical. Cities should involve residents in decision-making processes through participatory workshops, public consultations, and co-creation sessions (Leino and Puumala 2021). Additionally, targeted education campaigns, such as demonstrations of new technologies or success stories from similar projects, can address fears and foster a sense of shared ownership. For example, Medellín's cable car system succeeded partly due to its extensive community engagement efforts, which helped residents see its direct benefits (Galvin and Maassen 2020).

Smart cities represent a paradigm shift in urban planning by seamlessly integrating sustainability, resilience, and equity. By leveraging technology to balance environmental goals, social inclusion, and adaptive capacity, cities can create vibrant, liveable environments for all residents. Through a holistic and people-centred approach, smart cities can thrive as inclusive ecosystems that meet the challenges of today while preparing for a sustainable future.

## 5.4 Smart 15-Minute City

The integration of **smart city technologies** with the **15-minute city** concept marks a significant advancement in urban planning (Allam et al. 2022c). By combining localized accessibility with cutting-edge technological solutions, this model addresses modern urban challenges such as climate change, population growth, and social inequity.

Smart 15-minute cities leverage **real-time data**, **intelligent systems**, and **community-driven innovation** to create urban environments that are efficient, sustainable, and adaptable to residents' needs (Allam et al. 2022a). This section explores how technology enhances the fundamental principles of the 15-minute city, accessibility, sustainability, and inclusivity, while providing insights into its potential challenges and strategies for implementation.

## 5.5 Smart Mobility

Smart mobility is at the core of the Smart 15-Minute City, enabling residents to move efficiently impact within compact urban areas while reducing environmental and reliance on private cars (Jain et al. 2023).

Key innovations include:

- **Real-Time Transportation Systems**: Smart traffic lights, dynamic parking management, and real-time public transit updates improve traffic flow and reduce congestion. For example, smart traffic management system dynamically adjusts signals based on traffic density, significantly decreasing travel times and emissions (Ahmed et al. 2024).
- **Shared and Active Mobility**: Technologies supporting bike-sharing, e-scooters, and car-sharing platforms promote sustainable last-mile connectivity. For instance, the city of Barcelona integrated shared mobility solutions into their 15-minute city frameworks, encouraging active modes of transportation such as walking and cycling (Graells-Garrido et al. 2021).
- **Autonomous and Electric Vehicles**: Smart cities are transitioning to electric and autonomous public transit to achieve energy efficiency and accessibility. For instance, Singapore's autonomous bus system reduces emissions while providing seamless connectivity across neighbourhoods (Pathak et al. 2021).

These innovations allow the smart 15 minute cities achieve reduced travel times, lower emissions, and improved quality of life, aligning transportation systems with sustainability and accessibility goals.

## 5.6 Intelligent Land-Use Planning

**Intelligent land-use planning**, powered by technologies such as GIS and data analytics, ensures that urban spaces are designed to maximize convenience, efficiency, and adaptability (Mansourihanis et al. 2023). Intelligent land-use planning enhances the spatial design of 15-minute cities using data analytics to identify optimal locations for housing, businesses, and essential services (Fredriksson 2017).

Applications include:

- **Evidence-Based Spatial Design**: GIS tools analyse demographic, environmental, and mobility data to optimize the location of essential services like schools, hospitals, and retail hubs. This ensures equitable access to amenities while reducing travel distances (Zhand et al. 2024).
- **Mixed-Use and Walkable Communities**: Data-driven planning supports the creation of mixed-use neighbourhoods, where residential, commercial, and recreational spaces coexist. For example, Dutch land-use policies incorporate collaborative practices to adapt urban

layouts to societal changes, ensuring relevance and efficiency over time (Halleux et al. 2012).
- **Adaptable Urban Design**: Planners can reconfigure spaces based on shifting population needs, using real-time data to identify underutilized areas and reallocate resources (Rauws and Roo 2016; Dąbrowska-Żółtak et al. 2021).

By leveraging technology for land-use optimization, smart 15-minute cities promote **walkability**, **reduce urban sprawl**, and create vibrant, well-connected communities.

## 5.7 Robust Digital Infrastructure

A **robust digital infrastructure** is essential for integrating smart city technologies into the 15-minute city framework. High-speed internet, IoT networks, and cloud-based systems enable real-time communication and data sharing across urban systems (Rani and Chauhdary 2018). In addition, implementing a strong policy of infrastructure digitalization can lay the foundation for smart city innovations, boost economic development, and improve the quality of life of residents (Yeh 2017; Ismagilova et al. 2019; Shi et al. 2023).

Key applications include:

- IoT-Enabled Urban Services: Sensors deployed across the city monitor air quality, traffic flow, and waste management, enabling real-time adjustments (Zaman et al. 2024). In the context of a 15-minute city, where reducing reliance on cars is crucial, technologies such as smart traffic lights, dynamic parking pricing, and real-time public transit information make it easier for residents to travel short distances without unnecessary delays (Oladimeji et al. 2023).
- Smart Energy Grids: These grids integrate renewable energy sources, ensuring stable energy distribution (Meng 2023).
- Cybersecurity and Data Privacy: A critical component of digital infrastructure is ensuring the safety and privacy of urban data. Cities must implement data governance frameworks to protect resident information while leveraging it for urban innovation (Kalinin et al. 2021).

Digital infrastructure forms the foundation of smart 15-minute cities, ensuring that urban systems are efficient, responsive, and secure.

## 5.8 Sustainable Construction

Sustainable construction and building practices are vital for reducing the environmental impact of urban development in smart 15-minute cities. These practices further reinforce the environmental goals of the 15-minute city by incorporating green technologies into the urban fabric (Singla 2024).

Key practices include:

- Energy-Efficient Buildings: Smart buildings equipped with automated energy management systems optimize heating, cooling, and lighting (Iluyomade and Okwandu 2024). The Edge in Amsterdam, for example, uses AI-driven systems to minimize energy consumption (Grijp et al. 2019).
- Renewable Integration: Incorporating solar panels, wind turbines, and green roofs into urban infrastructure reduces reliance on non-renewable energy sources (Hoang et al. 2021).
- Eco-friendly Materials: Smart cities prioritize the use of sustainable construction materials, such as recycled concrete or timber, to minimize their carbon footprint (Nicolay et al. 2023).

Sustainable building practices align with the **climate goals** of the 15-minute city, ensuring that urban spaces are both energy-efficient and environmentally resilient.

## 5.9 Community Engagement

**Community engagement** ensures that the development of smart 15-minute cities reflects the diverse needs of residents (Allam et al. 2022a).

Digital platforms facilitate direct interaction between residents and city planners, allowing all groups in a community to contribute actively to urban development (Shrestha et al. 2024).

For instance, in Dutch countries, the Netherlands, Belgium, and Poland, land-use planning efficiently adapts regulatory instruments and collaborative practices to societal changes (Halleux et al. 2012).

Key strategies include:

- Participatory Platforms: Tools like CitizenLab allow residents to propose ideas, vote on initiatives, and co-create policies, fostering transparency and inclusion (Van Aeken 2017).
- Digital Literacy Programs: Providing education and training ensures that marginalized groups can fully participate in smart city initiatives, bridging the digital divide (Isabella and Agustian 2023).
- Feedback Mechanisms: Platforms like FixMyStreet enable residents to report issues, ensuring that urban services are responsive to community needs (Matthews et al. 2023).

By prioritizing community engagement, smart 15-minute cities ensure that technological advancements are inclusive and aligned with residents' aspirations.

## 5.10 Conclusion

The ***Smart 15-Minute City*** represents an innovative synthesis of two transformative urban planning models: the localized accessibility of the 15-minute city and the technological efficiency of the smart city. Together, they create a framework for cities that are sustainable, resilient, inclusive, and equitable, addressing the complex challenges of modern urbanization, such as climate change, social inequality, and rapid population growth. By blending human-centred urban design with cutting-edge digital systems, this model offers a holistic approach to shaping urban environments that prioritize both technological advancement and the well-being of all residents.

At its core, the Smart 15-Minute City enhances the liveability of neighbourhoods through ***smart mobility***, ***intelligent land-use planning***, and ***sustainable infrastructure & building construction practices*** (Fig. 5.2). For instance, technologies like real-time traffic management, dynamic public transit systems, and IoT-enabled services reduce congestion, shorten travel times, and encourage active transportation options such as walking and cycling. Meanwhile, data-driven spatial planning ensures that essential services, such as schools, healthcare facilities, and recreational spaces, are equitably distributed across compact urban areas. Moreover, sustainable building practices, including energy-efficient materials and renewable energy systems, strengthen the model's environmental focus, ensuring urban resilience against future challenges. These innovations not only reduce urban sprawl and reliance on private vehicles but also minimize environmental impacts, aligning cities with global sustainability goals.

The integration of **smart digital infrastructure** further amplifies the 15-minute city model by enabling cities to become dynamic and responsive ecosystems. Through real-time data collection, predictive analytics, and participatory platforms, urban planners can optimize resource allocation, enhance public services, and empower residents to shape their environments.

Equity is another central tenet of the Smart 15-Minute City, ensuring that technological advancements do not exacerbate existing disparities. By addressing digital divides, expanding access to affordable connectivity, and promoting digital literacy, cities can ensure that marginalized communities are not excluded from the benefits of urban innovation. Equally, privacy safeguards and robust data governance frameworks are essential to maintaining public trust and fostering widespread participation.

Despite its vast potential, the Smart 15-Minute City also faces significant challenges, including **financial constraints**, **privacy concerns**, and **resistance to change**. However, these obstacles can be mitigated through strategies such as PPPs, targeted educational initiatives, and transparent community engagement. Successful examples,

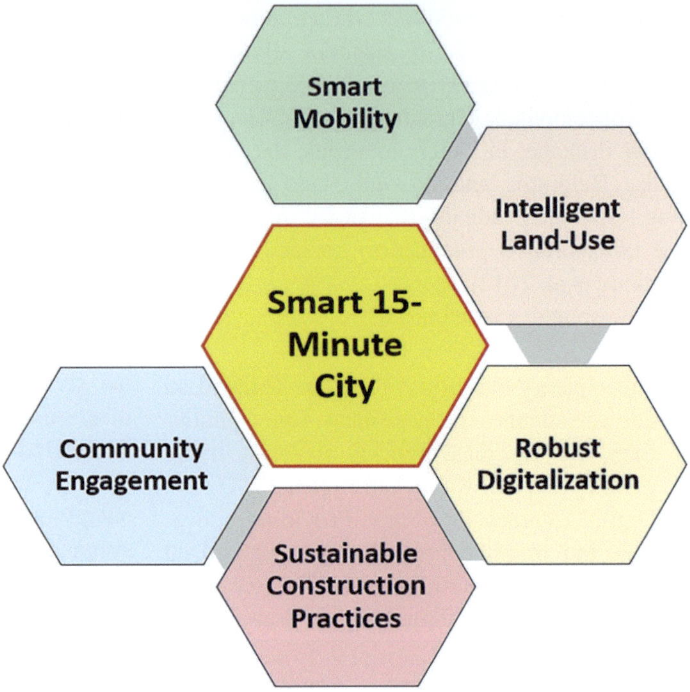


**Fig. 5.2** Components of a smart 15-minute city

such as Medellín's integration of cable cars with public transit to connect marginalized communities, demonstrate how innovative and inclusive approaches can transform urban spaces.

In conclusion, the **Smart 15-Minute City** offers a compelling vision for sustainable and equitable urban development by merging localized urban design with technological innovation. This integrated model allows cities to not only adapt to the demands of modern life but also thrive in the face of future challenges. By focusing on **proximity, efficiency, and inclusivity**, smart 15-minute cities foster vibrant neighbourhoods where residents can access essential services, participate in decision-making, and enjoy a high quality of life, all while reducing environmental impacts and promoting social equity. For policymakers and urban planners, embracing this hybrid framework represents a critical opportunity to build cities that are not only technologically advanced but also resilient, human-centred, and capable of addressing the diverse needs of their populations.

## References

Adreani L, Bellini P, Fanfani M et al (2024) Smart city digital twin framework for real-time multi-data integration and wide public distribution. IEEE Access 12:76277–76303. https://doi.org/10.1109/ACCESS.2024.3406795

Ai J (2024) Revolutionizing urban waste management in San Francisco: the role of technology-driven solutions in advancing circular economy practices. Int J Bus Technol Manag. https://doi.org/10.55057/ijbtm.2024.6.1.41

Ahamer G (2021) International financial institutions ask to contribute to climate protection. Financ Theory Pract 25:6–23. https://doi.org/10.26794/2587-5671-2020-25-4-6-23

Ahmed R, Osman RA, Amer M (2024) Navigating urban congestion: a comprehensive strategy based on an efficient smart IoT wireless communication for PV powered smart traffic management system. PLoS ONE 19:e0310002. https://doi.org/10.1371/journal.pone.0310002

Allam Z, Bibri SE, Chabaud D, Moreno C (2022a) The theoretical, practical, and technological foundations of the 15-minute city model: proximity and its environmental, social and economic benefits for sustainability. Energies 15:6042. https://doi.org/10.3390/en15166042

Allam Z, Bibri SE, Chabaud D, Moreno C (2022b) The '15-Minute City' concept can shape a net-zero urban future. Humanit Soc Sci Commun 9:126. https://doi.org/10.1057/s41599-022-01145-0

Allam Z, Bibri SE, Jones DS et al (2022c) Unpacking the '15-Minute City' via 6G, IoT, and digital twins: towards a new narrative for increasing urban efficiency, resilience, and sustainability. Sensors 22:1369. https://doi.org/10.3390/s22041369

Ammar M, Utomo S, Sarminingsih A (2024) Harmonizing urban futures: integrating smart and sustainable city principles. Int J Innov Sci Res Technol 418–430. https://doi.org/10.38124/ijisrt/IJISRT24SEP083

Bano S, Hussain A, Arif A et al (2024) Data-driven Internet of Things: role in smart cities. Asian Bull Big Data Manag 4. https://doi.org/10.62019/abbdm.v4i02.155

Bastos D, Fernández-Caballero A, Pereira A, Rocha NP (2022) Smart city applications to promote citizen participation in city management and governance: a systematic review. Informatics 9:89. https://doi.org/10.3390/informatics9040089

Bifulco F, Tregua M, Amitrano CC, D'Auria A (2016) ICT and sustainability in smart cities management. Int J Public Sect Manag 29:132–147. https://doi.org/10.1108/IJPSM-07-2015-0132

Borge R, Balcells J, Padró-Solanet A (2023) Democratic disruption or continuity? Analysis of the Decidim platform in Catalan municipalities. Am Behav Sci 67:926–939. https://doi.org/10.1177/00027642221092798

Brand P, Dávila JD (2011) Mobility innovation at the urban margins. City 15:647–661. https://doi.org/10.1080/13604813.2011.609007

Braun T, Fung BCM, Iqbal F, Shah B (2018) Security and privacy challenges in smart cities. Sustain Cities Soc 39:499–507. https://doi.org/10.1016/j.scs.2018.02.039

Cesario E (2023) Big data analytics and smart cities: applications, challenges, and opportunities. Front Big Data 6:119402. https://doi.org/10.3389/fdata.2023.1149402

Cerrillo-Martínez A, Casadesús-de-Mingo A (2021) Data governance for public transparency. El Prof La Inf. https://doi.org/10.3145/epi.2021.jul.02

Chai Z, Shah SA, Draheim D et al (2024) Guest editorial: smart cities 2.0: how artificial intelligence and Internet of Things are transforming urban living. IET Smart Cities 6:129–131. https://doi.org/10.1049/smc2.12091

Choudhary H, Bansal N (2022) Addressing Digital Divide through Digital Literacy Training Programs: A Systematic Literature Review. Digit Educ Rev 224–248. https://doi.org/10.1344/der.2022.41.224-248

Dąbrowska-Żółtak K, Wojtowicz J, Wrona S (2021) Reconfigurable Neighborhood—Mechatronisation of the Urban Design. Sustainability 13:13547. https://doi.org/10.3390/su132413547

DuBose BM, Mayo AM (2020) Resistance to change: a concept analysis. Nurs Forum 55:631–636. https://doi.org/10.1111/nuf.12479

Ertiö T-P, Tuominen P, Rask M (2019) Turning ideas into proposals: a case for blended participation during the participatory budgeting trial in Helsinki, pp 15–25

Fainstein SS (2018) Resilience and justice: planning for New York City. Urban Geogr 39:1268–1275. https://doi.org/10.1080/02723638.2018.1448571

Filazzola A, Shrestha N, MacIvor JS (2019) The contribution of constructed green infrastructure to urban biodiversity: A synthesis and meta-analysis. J Appl Ecol 56:2131–2143. https://doi.org/10.1111/1365-2664.13475

Fredriksson A (2017) Location-allocation of public services—Citizen access, transparency and measurement. A method and evidence from Brazil and Sweden. Socioecon Plann Sci 59:1–12. https://doi.org/10.1016/j.seps.2016.09.008

Galvin M, Maassen A (2020) Connecting formal and informal spaces: a long-term and multi-level view of Medellín's Metrocable. Urban Transform 2:4. https://doi.org/10.1186/s42854-020-00008-8

Gorges G de C, Wedekind JP, Kimura GAP et al (2020) Generation of renewable energy through roads, bridges and cycle paths. Rev Mundi Eng Tecnol e Gestão 5. https://doi.org/10.21575/25254782rmetg2020vol5n71098. ISSN 2525–4782

Gil-Rivas V, Kilmer RP (2016) Building community capacity and fostering disaster resilience. J Clin Psychol 72:1318–1332. https://doi.org/10.1002/jclp.22281

Graells-Garrido E, Serra-Burriel F, Rowe F et al (2021) A city of cities: measuring how 15-minutes urban accessibility shapes human mobility in Barcelona. PLoS ONE 16:e0250080. https://doi.org/10.1371/journal.pone.0250080

Golub A, Satterfield V, Serritella M et al (2019) Assessing the barriers to equity in smart mobility systems: a case study of Portland, Oregon. Case Stud Transp Policy 7:689–697. https://doi.org/10.1016/j.cstp.2019.10.002

Hadiyana T, Ji-hoon S (2024) AI-driven urban planning: enhancing efficiency and sustainability in smart cities. ITEJ (Inf Technol Eng J) 9:23–35. https://doi.org/10.24235/itej.v9i2.124

Halleux J-M, Marcinczak S, van der Krabben E (2012) The adaptive efficiency of land use planning measured by the control of urban sprawl. The cases of the Netherlands, Belgium and Poland. Land Use Policy 29:887–898. https://doi.org/10.1016/j.landusepol.2012.01.008

Hoang AT, Pham VV, Nguyen XP (2021) Integrating renewable sources into energy system for smart city as a sagacious strategy towards clean and sustainable process. J Clean Prod 305:127161. https://doi.org/10.1016/j.jclepro.2021.127161

Iluyomade TD, Okwandu AC (2024) Smart buildings and sustainable design: leveraging AI for energy optimization in the built environment. Int J Sci Res Arch 12:2448–2456. https://doi.org/10.30574/ijsra.2024.12.1.1049

Isabella I, Agustian E (2023) Implementing digital literacy policies and the challenges of towards smart city in Palembang City. J Gov Local Polit 5:122–132. https://doi.org/10.47650/jglp.v5i2.936

Ismagilova E, Hughes L, Dwivedi YK, Raman KR (2019) Smart cities: advances in research—An information systems perspective. Int J Inf Manage 47:88–100. https://doi.org/10.1016/j.ijinfomgt.2019.01.004

Jain R, Dhingra S, Joshi K et al (2023) Enhance traffic flow prediction with real-time vehicle data integration. J Auton Intell 6:574. https://doi.org/10.32629/jai.v6i2.574

Kalair A, Abas N, Saleem MS et al (2021) Role of energy storage systems in energy transition from fossil fuels to renewables. Energy Storage 3:135. https://doi.org/10.1002/est2.135

Kalinin M, Krundyshev V, Zegzhda P (2021) Cybersecurity risk assessment in smart city infrastructures. Machines 9:78. https://doi.org/10.3390/machines9040078

Kamel Boulos MN, Al-Shorbaji NM (2014) On the Internet of Things, smart cities and the WHO healthy cities. Int J Health Geogr 13:10. https://doi.org/10.1186/1476-072X-13-10

King SF, Brown P (2007) Fix my street or else. In: Proceedings of the 1st international conference on theory and practice of electronic governance. ACM, New York, NY, USA, pp 72–80

Kolotouchkina O, Ripoll González L, Belabas W (2024) Smart cities, digital inequalities, and the challenge of inclusion. Smart Cities 7:3355–3370. https://doi.org/10.3390/smartcities7060130

Laffta S, Al-rawi A (2018) Green technologies in sustainable urban planning. MATEC Web Conf 162:05029. https://doi.org/10.1051/matecconf/201816205029

Leach M, Reyers B, Bai X et al (2018) Equity and sustainability in the anthropocene: a social–ecological systems perspective on their intertwined futures. Glob Sustain 1:e13. https://doi.org/10.1017/sus.2018.12

Lee J, Babcock J, Pham TS et al (2023) Smart city as a social transition towards inclusive development through technology: a tale of four smart cities. Int J Urban Sci 27:75–100. https://doi.org/10.1080/12265934.2022.2074076

Leino H, Puumala E (2021) What can co-creation do for the citizens? applying co-creation for the promotion of participation in cities. Environ Plan C Polit Sp 39:781–799. https://doi.org/10.1177/2399654420957337

Lv Z, Hu B, Lv H (2020) Infrastructure monitoring and operation for smart cities based on IoT system. IEEE Trans Ind Informatics 16:1957–1962. https://doi.org/10.1109/TII.2019.2913535

Mansourihanis O, Maghsoodi Tilaki MJ, Yousefian S, Zaroujtaghi A (2023) A computational geospatial approach to assessing land-use compatibility in urban planning. Land 12:2083. https://doi.org/10.3390/land12112083

Marchigiani E (2023) Building accessible cities: A reflection through time, towards future perspectives

Matheus R, Janssen M (2020) A systematic literature study to unravel transparency enabled by open government data: the window theory. Public Perform Manag Rev 43:503–534. https://doi.org/10.1080/15309576.2019.1691025

Matthews P, Parsons A, Nyanzu E, Rae A (2023) Dog fouling and potholes: understanding the role of coproducing ‘citizen sensors’ in local governance. Local Gov Stud 49:908–931. https://doi.org/10.1080/03003930.2022.2116575

Meng W (2023) Analysis of the role of smart grids in renewable energy. MATEC Web Conf 386:03013. https://doi.org/10.1051/matecconf/202338603013

Mottahedi A, Sereshki F, Ataei M et al (2021) The Resilience of critical infrastructure systems: a systematic literature review. Energies 14:1571. https://doi.org/10.3390/en14061571

Moreno C, Allam Z, Chabaud D et al (2021) Introducing the “15-Minute City”: sustainability, resilience and place identity in future post-pandemic cities. Smart Cities 4:93–111. https://doi.org/10.3390/smartcities4010006

Narain S (2024) Integrating smart city technologies for enhanced urban sustainability. J Sustain Solut 1:13–17. https://doi.org/10.36676/j.sust.sol.v1.i3.15

Nakamura H, Oosawa M (2021) Effects of the underground discharge channel/reservoir for small urban rivers in the Tokyo area. IOP Conf Ser Earth Environ Sci 703:012029. https://doi.org/10.1088/1755-1315/703/1/012029

Nicolay P, Schlögl S, Thaler SM et al (2023) Smart materials for green(er) cities, a short review. Appl Sci 13:9289. https://doi.org/10.3390/app13169289

Nižetić S, Djilali N, Papadopoulos A, Rodrigues JJPC (2019) Smart technologies for promotion of energy efficiency, utilization of sustainable resources and waste management. J Clean Prod 231:565–591. https://doi.org/10.1016/j.jclepro.2019.04.397

Oladimeji D, Gupta K, Kose NA et al (2023) Smart transportation: an overview of technologies and applications. Sensors 23:3880. https://doi.org/10.3390/s23083880

Pathak A, Sethuraman G, Ongel A, Lienkamp M (2021) Impacts of electrification & automation of public bus transportation on sustainability—A case study in Singapore. Forsch Im Ingenieurwes 85:431–442. https://doi.org/10.1007/s10010-020-00408-z

Pandey A, Pandey A, Saini JR (2024) Decentralized energy markets enhancing peer-to-peer energy sharing with blockchain. In: 2024 5th International conference on smart electronics and communication (ICOSEC). IEEE, pp 874–881

Patel S (2024) Perspective of digital transformation in rural India. Int J Sci Res 13:387–390. https://doi.org/10.21275/SR24402140819

Qiu D, Strbac G, Wang Y et al (2024) Artificial intelligence for microgrid resilience: a data-driven and model-free approach. IEEE Power Energy Mag 22:18–27. https://doi.org/10.1109/MPE.2024.3405893

Radchenko K (2022) Public-private partnership for smarter cities. Cent East Eur eDem eGov Days 341:103–114. https://doi.org/10.24989/ocg.v341.7

Ramirez Lopez LJ, Grijalba Castro AI (2020) Sustainability and resilience in smart city planning: a review. Sustainability 13:181. https://doi.org/10.3390/su13010181

Rauws W, De Roo G (2016) Adaptive planning: Generating conditions for urban adaptability. Lessons from Dutch organic development strategies. Environ Plan B Plan des 43:1052–1074. https://doi.org/10.1177/0265813516658886

Rani S, Chauhdary SH (2018) A novel framework and enhanced QoS big data protocol for smart city applications. Sensors 18:3980. https://doi.org/10.3390/s18113980

Rendon C, Osman KK, Faust KM (2021) Path towards community resilience: examining stakeholders' coordination at the intersection of the built, natural, and social systems. Sustain Cities Soc 68:102774. https://doi.org/10.1016/j.scs.2021.102774

Roostaie S, Nawari N, Kibert CJ (2019) Sustainability and resilience: a review of definitions, relationships, and their integration into a combined building assessment framework. Build Environ 154:132–144. https://doi.org/10.1016/j.buildenv.2019.02.042

Sanguino R, Barroso A, Fernández-Rodríguez S, Sánchez-Hernández MI (2020) Current trends in economy, sustainable development, and energy: a circular economy view. Environ Sci Pollut Res 27:1–7. https://doi.org/10.1007/s11356-019-07074-x

Sha K, Taeihagh A, De Jong M (2024) Governing disruptive technologies for inclusive development in cities: a systematic literature review. Technol Forecast Soc Change 203:123382. https://doi.org/10.1016/j.techfore.2024.123382

Shrestha R, Hasselder P, Bolte G (2024) Digitally supported participation in the nexus between public health and urban planning. Bundesgesundheitsblatt - Gesundheitsforsch - Gesundheitsschutz 67:316–323. https://doi.org/10.1007/s00103-024-03838-0

Shi Y, Zhang T, Jiang Y (2023) Digital economy technological innovation and urban resilience. Sustainability 15:9250. https://doi.org/10.3390/su15129250

Singla A (2024) Sustainable urban infrastructure: innovations in green building technologies and urban planning. Univers Res Reports 11:106–112. https://doi.org/10.36676/urr.v11.i4.1316

Sousa MJ (2023) Blockchain as a driver for transformations in the public sector. Policy des Pract 6:415–432. https://doi.org/10.1080/25741292.2023.2267864

Sultana F, Bhardwaj R, Kumar RN, et al (2024) Smart traffic management system for efficient mobility and emergency response. In: 2024 international conference on knowledge engineering and communication systems (ICKECS). IEEE, pp 1–5

Suárez G (2022) The seismic early warning system of Mexico (SASMEX): a retrospective view and future challenges. Front Earth Sci 10:827236. https://doi.org/10.3389/feart.2022.827236

Udo WS, Kwakye JM, Ekechukwu DE, Ogundipe OB (2023) Predictive analytics for enhancing solar energy forecasting and grid integration. Eng Sci Technol J 4:589–602. https://doi.org/10.51594/estj.v4i6.1394

Van Aeken K (2016) Digital democracy in belgium and the Netherlands. A socio-legal analysis of Citizenlab.be and Consultatie.nl. SSRN Electron J https://doi.org/10.2139/ssrn.2932962

Vassilakopoulou P, Hustad E (2023) Bridging digital divides: a literature review and research agenda for information systems research. Inf Syst Front 25:955–969. https://doi.org/10.1007/s10796-020-10096-3

van Aeken K (2017) Digital democracy in Belgium and the Netherlands. A socio-legal analysis of technologies, embedding and expectations of two fourth wave innovations. In: Digital Democracy in a globalized world. Edward Elgar Publishing

van der Grijp N, van der Woerd F, Gaiddon B et al (2019) Demonstration projects of nearly zero energy buildings: lessons from end-user experiences in Amsterdam, Helsingborg, and Lyon. Energy Res Soc Sci 49:10–15. https://doi.org/10.1016/j.erss.2018.10.006

Wang QR (2024) Towards zero-emission urban mobility: leveraging AI and LCA for targeted interventions. Build Simul 17:1653–1657. https://doi.org/10.1007/s12273-024-1193-7

Wu M, Yan B, Huang Y, Sarker MNI (2022) Big data-driven urban management: potential for urban sustainability. Land 11:680. https://doi.org/10.3390/land11050680

Wu Q, Huang Y, Irga P et al (2024) Synergistic control of urban heat island and urban pollution island effects using green infrastructure. J Environ Manage 370:122985. https://doi.org/10.1016/j.jenvman.2024.122985

Yeh H (2017) The effects of successful ICT-based smart city services: From citizens' perspectives. Gov Inf Q 34:556–565. https://doi.org/10.1016/j.giq.2017.05.001

Yu M (2024) Key challenges and potential solutions in smart city governance. Sci Soc Res 6:173–180. https://doi.org/10.26689/ssr.v6i8.7818

Zaman M, Puryear N, Abdelwahed S, Zohrabi N (2024) A review of IoT-based smart city development and management. Smart Cities 7:1462–1501. https://doi.org/10.3390/smartcities7030061

Zhand S, Simons G, Karimi K, et al (2024) Evidence-based design and planning; reflections from past and current theories and practices

# Applications of the 15-Minute City Model 6

## 6.1 Introduction

The 15-minute city concept has emerged as a ground-breaking approach to urban planning, addressing the dual challenges of urbanization and climate change by promoting accessibility, sustainability, and equity (Moreno et al. 2021). By reducing car dependency, encouraging localized living, and fostering greener neighbourhoods, the 15-minute city aims to transform urban life (Allam et al. 2022b; Moreno et al. 2024), while aligning with the United Nations Sustainable Development Goals (SDGs), as depicted in the Fig. 6.1.

Recent research by Bencekri et al. (2025) highlighted the environmental potential of the 15-minute city but also underscores critical gaps in current understanding. Their hybrid methodology, including systematic literature review, topic modelling, and sentiment analysis, reveals that only 10.2% of existing studies focus on environmental outcomes, with terms like ***emissions*** and ***decarbonization*** often missing from the discourse. While sentiment analysis reflects cautious optimism about the model's ability to address climate challenges, concerns remain regarding its real-world effectiveness. The study called for more empirical research to quantify the 15-minute city's environmental impact, particularly in reducing emissions and enhancing urban resilience. These findings emphasize the need for a more evidence-driven approach to integrating the 15-minute city model into sustainable urban development.

Building on these insights, this chapter explores the diverse applications of the 15-minute city across several key dimensions of urban planning and development. By integrating localized design principles with advanced urban planning tools and strategies, the 15-minute city serves as a practical framework for addressing a variety of urban challenges:

- **Decarbonization and Climate Action:** The 15-minute city fosters localized living and sustainable transportation options, reducing greenhouse gas emissions and mitigating climate change. This section explores how the model promotes energy-efficient neighbourhoods and low-carbon mobility solutions.
- **Planning Support Systems (PSS) for Active Mobility:** Advanced tools such as PSS leverage data analytics and spatial modelling to design walkable and bikeable cities. By prioritizing active transportation, these systems enable planners to enhance accessibility and reduce car dependency.
- **Mobility Hubs and Seamless Connectivity:** Mobility hubs act as central nodes within the 15-minute city, integrating walking, cycling, public transit, and shared mobility services. These hubs streamline travel, enhance connectivity, and reduce reliance on private vehicles, particularly in dense urban areas.



M. Bencekri and S. Lee, *Sustainable Urban Evolution*, Sustainable Development Goals Series,
https://doi.org/10.1007/978-981-96-5146-7_6

**Fig. 6.1** 15-minute city alignment with SDGs

- **Economic Revitalization and Local Growth:** By fostering vibrant, localized economies, the 15-minute city model reduces commuting times, attracts investment, and creates more resilient urban economies. This approach not only improves productivity but also enhances residents' quality of life.
- **Equity and Social Inclusion:** A fundamental principle of the 15-minute city is ensuring equitable access to essential services for all residents, regardless of socioeconomic status. This section discusses how the model addresses systemic disparities by creating balanced, inclusive urban environments that provide equal opportunities for all.

Through these applications, this chapter offers a comprehensive exploration of how the 15-minute city model can be implemented to address pressing urban sustainability challenges while improving the well-being of residents. By aligning environmental sustainability, social equity, and economic vitality, the 15-minute city represents a holistic and forward-thinking blueprint for the cities of tomorrow.

## 6.2 Decarbonization in a 15-Minute City

### 6.2.1 Introduction

The 15-minute city model addresses the dual challenges of urban sprawl and carbon emissions by encouraging more localized living. As Moreno et al. (2021) note, this model not only reduces the need for motorized transportation, but also promotes public health and community engagement by fostering active modes of transport. By ensuring that residents can access workplaces, schools, healthcare, and leisure activities within a short distance, the 15-minute city significantly reduced the environmental impact associated with daily commutes (Fig. 6.2).

Jin et al. (2024) examined 12 American cities and highlighted the potential of the 15-minute city concept to reduce $CO_2$ emissions by promoting proximity-based planning, which can decrease reliance on cars and encourage low-emission transportation modes, such as walking, cycling, and public transit. The study found that despite the current reliance on cars, there is substantial poten-

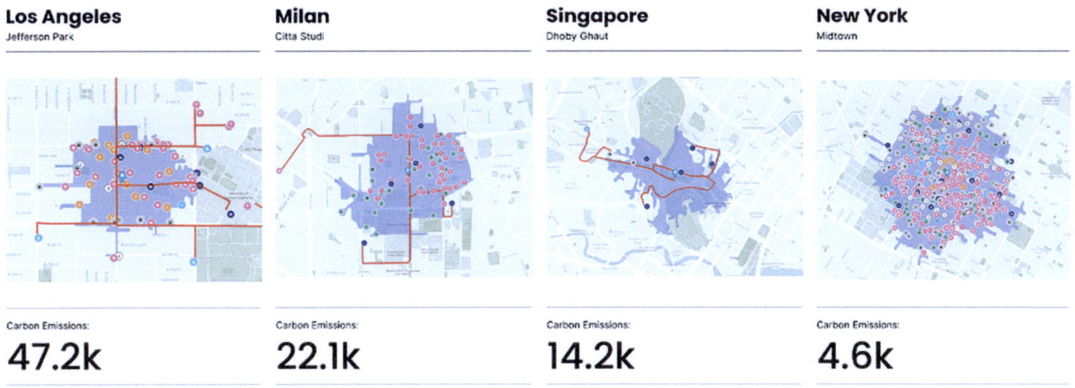


**Fig. 6.2** Proximity benefits of reduced carbon (Kondratenko 2023)

tial for realizing a 15-minute city vision. Most residents' daily activities could be accommodated within a 15-minute radius using low-emission transportation modes.

While the 15-minute city model has been widely discussed for its potential to enhance urban sustainability and liveability, a key research gap exists regarding its specific environmental impacts. Previous studies have primarily addressed the model in broad sustainability terms, with limited attention to its ability to address environmental challenges such as greenhouse gas (GHG) emissions, urban sprawl, and ecosystem degradation (Bencekri et al. 2025).

### 6.2.2 Methodology

Bencekri et al. (2025) employed a hybrid methodology combining a systematic literature review (guided by PRISMA protocols) with advanced machine learning techniques, including topic modelling and sentiment analysis, to explore the 15-minute city model's potential for mitigating environmental impacts such as GHG emissions, urban sprawl, and ecological degradation (Fig. 6.3).

#### 6.2.2.1 Systematic Review (PRISMA-Guided)

A systematic review was conducted to ensure a high-quality, unbiased dataset. This involved:

- Searching three academic databases (Web of Science, Scopus, Scispace), yielding 3,160 initial records.
- Screening studies based on keywords such as ***sustainability, pollution, and emissions,*** as well as their relevance to the environmental impacts of the model.
- Filtering and assessing 169 peer-reviewed studies that directly addressed the research question. ***PRISMA*** protocols ensured transparency and minimized selection bias.
- Synthesizing insights from qualitative and quantitative studies, with a focus on the 15-minute city model's ability to reduce carbon emissions, mitigate environmental degradation, and enhance urban sustainability.

#### 6.2.2.2 Topic Modelling

Topic modelling was employed to uncover latent themes and trends in the selected body of literature. The process involved:

- Preprocessing text (titles, abstracts, and keywords) by tokenizing words, removing stop words, and constructing a dictionary of terms.
- Applying **LDA**, a probabilistic machine learning model, to cluster terms into topics based on co-occurrence patterns.
- Enhancing the analysis with a **hybrid LDA and word embedding model**, which contextualized relationships among terms and improved topic coherence.

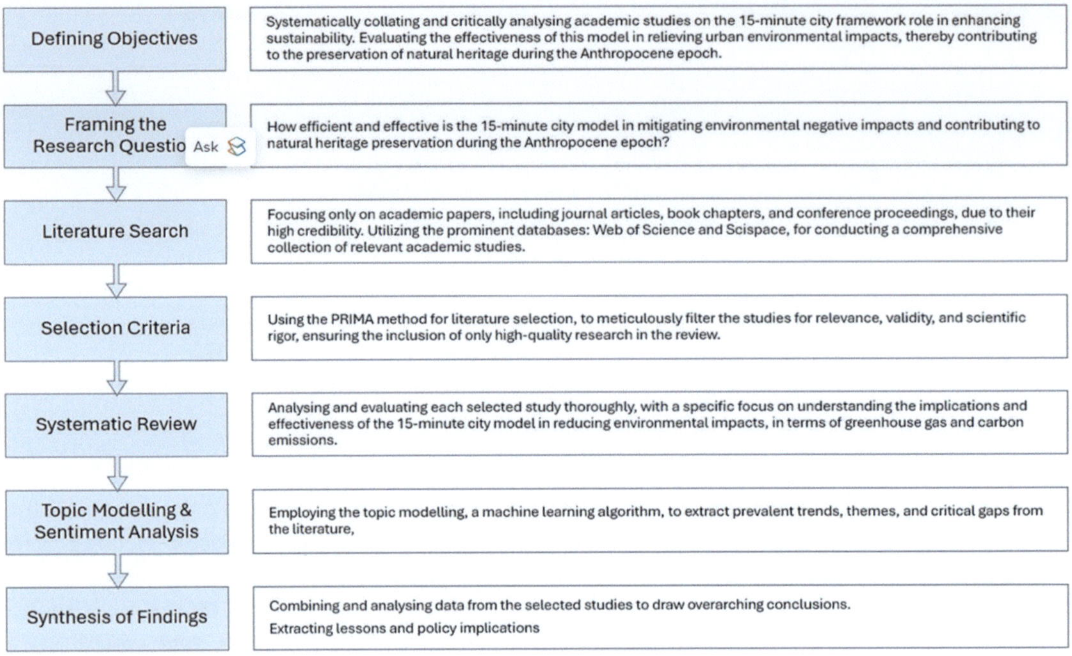


**Fig. 6.3** Methodology to explore 15-minute city environmental impacts

- Visualizing results using tools such as t-SNE clustering and trend analysis for greater interpretability, revealing the underexplored areas in the literature.

#### 6.2.2.3 Sentiment Analysis

Sentiment analysis was applied to quantify the tone of academic discourse regarding the environmental impacts of the 15-minute city framework. This involved:

- Analysing abstracts, titles, and keywords using a rule-based polarity lexicon to assign sentiment scores ranging from -1 (negative) to + 1 (positive).
- Identifying whether scholars were optimistic, neutral, or skeptical about the model's feasibility and benefits.
- Providing an additional layer of insight, complementing topic modelling by highlighting the prevailing attitudes and emotional framing in the literature.

### 6.2.3 Findings Synthesis

#### 6.2.3.1 Thematic Analysis

A systematic review of 150 academic sources revealed three primary thematic groups in the literature, highlighting areas of focus and gaps in the study of the 15-minute city model:

- Environmental Impact (10.2% of Studies):
  - Focuses on decarbonization, emissions reduction, and ecological sustainability.
  - Key Insight: Although the model is theoretically linked to reducing emissions and promoting green urban environments, empirical evidence remains limited.
- Sustainability and Resilience:
  - Explores the alignment of the 15-minute city model with broader sustainability goals, including social equity, liveability, and resilience.
  - Key Insight: While this theme emphasizes sustainable urban development, it often

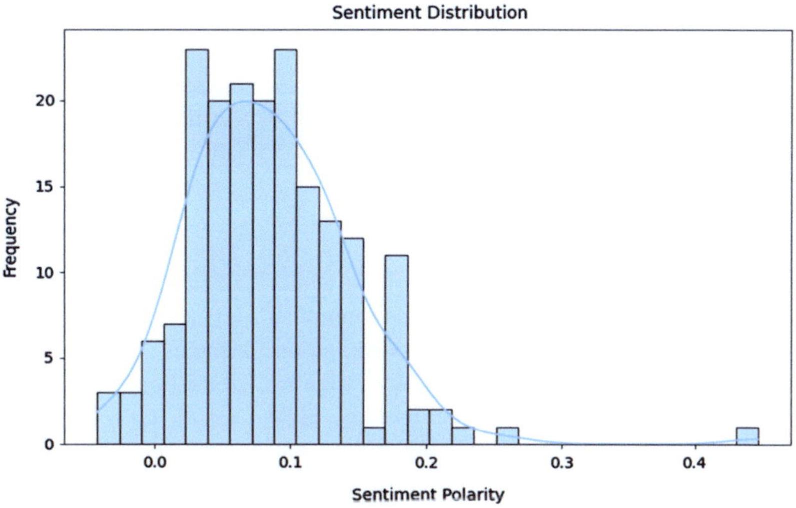


**Fig. 6.4** Sentiment analysis findings

lacks direct evaluations of environmental benefits specific to the model.

- Urban Planning and Accessibility:
  - Dominates the literature, emphasizing mobility, proximity, and inclusivity in urban design.
  - Key Insight: This theme underscores the role of the 15-minute city in improving liveability and urban mobility but frequently neglects its environmental contributions.

The disproportionate focus on urban planning and general sustainability overshadows detailed studies on the model's ability to mitigate environmental challenges, such as climate resilience and emissions reduction.

#### 6.2.3.2 Topic Modelling Analysis

Secondly, the study applied topic modelling (**LDA** and **Word2Vec**) to analyse the research corpus, uncovering five core topics:

- **Integrative Urban Planning Models**: Focus on mobility and accessibility in city planning.
- **Urban Service Efficiency**: Emphasis on optimizing infrastructure to minimize travel time.
- **Spatial Development in Urban Expansion**: Exploration of spatial planning in growing cities.
- **Theoretical Urban Space Utilization**: Examination of conceptual frameworks for urban planning.
- **Pragmatic Approaches to Urban Planning**: Analysis of real-world implementation challenges.

The key finding of this topic modelling is that the environmental terms like ***decarbonization*** and ***sustainability*** are underrepresented, underscoring a research bias towards urban design and mobility over environmental impacts.

#### 6.2.3.3 Sentiment Analysis

Lastly, using ***TextBlob***, sentiment analysis assessed the tone of academic discourse on the 15-minute city model (Fig. 6.4):

- **Neutral to Slightly Positive Sentiments**: Most papers (polarity scores between 0.0 and 0.2) express cautious optimism about the model's environmental potential.
- **Absence of Negative Sentiments**: No strongly critical perspectives were identified, reflecting a consensus on the model's promise but with recognition of implementation challenges.

Therefore, while scholars are generally optimistic about the 15-minute city, the neutral tone reflects the lack of robust empirical evidence for its environmental efficacy.

### 6.2.4 Discussions

The 15-minute city model aims to enhance urban sustainability by reducing car dependency and facilitating access to essential services within walking or cycling distances. This approach theoretically lowers emissions, reduces urban sprawl, and promotes green mobility, aligning closely with the UN-SDGs, particularly SDG 11 (Sustainable Cities and Communities) and SDG 13 (Climate Action). Scholars like Allam et al. (2022b) and Lima et al. (2022) have highlighted the potential of mixed land use and proximity (Marchigiani and Bonfantini 2022)tematic review of 150 academic sources revealed significant research gaps, with only 10.2% of the literature addressing the model's environmental contributions, such as its capacity to reduce emissions and enhance climate resilience.

Although the 15-minute city is widely regarded as a socially inclusive and mobility-focused framework, much of the current research remains conceptual, with insufficient empirical evidence to validate its environmental impact. Scholars like Barbieri et al. (2023) and Marchigiani and Bonfantini (2022) have acknowledged the theoretical appeal of the model but questioned its practical feasibility in cities with established infrastructure and its innovation, comparing it to older planning strategies like Garden Cities and New Urbanism, which often failed to deliver long-term environmental benefits. Casarin et al. (2023) have further emphasized the need for systematic studies to assess its direct contributions to emissions reduction and climate change mitigation.

In this study conducted by Bencekri et al. (2025), topic modelling identified five key themes: (1) Integrative Urban Planning Models, (2) Urban Service Efficiency, (3) Spatial Development in Urban Expansion, (4) Theoretical Urban Space Utilization, and (5) Pragmatic Approaches to Urban Planning. While these themes align with the 15-minute city's focus on accessibility, mobility, and equity, environmental sustainability remains underexplored. Keywords such as ***emissions, sustainable, and climate resilience*** were notably absent, indicating a narrow framing of the model's potential.

On the other hand, sentiment analysis of scholarly discourse further revealed neutral to slightly positive sentiments about the model's environmental promise (Bencekri et al. 2025). Most studies acknowledged the potential for emissions reduction, but scholars expressed cautious optimism, reflecting uncertainty about the model's practical implementation and long-term ecological contributions. There was no negative sentiment, indicating that the academic community views the model as promising but incomplete.

### 6.2.5 Conclusion

The findings highlight the need for future research to shift from theoretical discussions to empirical investigations. Specifically, studies should focus on quantifying the 15-minute city's environmental impact, such as measuring emissions reductions, energy efficiency gains, and contributions to climate resilience.

In addition, further exploration is needed to understand how the model can be adapted to cities with diverse geographic, cultural, and economic contexts. Also, researchers and planners should integrate environmental assessments with urban planning methodologies to create a more holistic evaluation of the model's contributions to sustainability.

Ultimately, while the 15-minute city model holds significant theoretical promise for addressing urban sustainability challenges, its success hinges on addressing the critical research gaps identified in this review. By providing robust empirical evidence of its environmental benefits, the model can advance beyond a conceptual ideal to become a practical, scalable solution for achieving SDG goals and creating inclusive, resilient cities in the Anthropocene epoch.

## 6.3 Mobility Hubs for a 15-Minute City

### 6.3.1 Introduction

A 15-minute city is a transformative urban planning concept designed to reduce carbon emissions by creating compact, self-sufficient neighbourhoods where essential services are accessible within a 15-minute walk or bike ride (Moreno et al. 2021). As cities grapple with the pressing challenges of climate change, this model offers a promising pathway for decarbonization by minimizing reliance on private vehicles and promoting sustainable transportation options. Central to this approach is the integration of carbon–neutral mobility hubs, which act as focal points for low-carbon transportation modes. This section explores the role of the 15-minute city in decarbonization, with a focus on innovative strategies for locating carbon–neutral mobility hubs (Bencekri et al. 2024a). By employing advanced artificial intelligence (AI) techniques, this study provides a framework for optimizing the placement of these hubs, thereby enhancing the overall sustainability and efficiency of urban mobility systems.

### 6.3.2 Mobility Hubs

Mobility hubs are becoming increasingly recognized as an essential policy tool for implementing a 15-minute city, particularly in dense urban areas where space and resources are limited (Bencekri et al. 2024a). A mobility hub, as shown in Fig. 6.5, is a central point where various forms of transportation, such as public transit, bike-sharing, electric scooters, and ride-sharing services converge, enabling seamless multimodal travel (Arnold et al. 2023). These hubs are designed to reduce reliance on private vehicles, promote sustainable transportation, and enhance connectivity in urban environments (Dilks 2021). In the context of the 15-minute city, mobility hubs play a crucial role in ensuring that residents have easy access to essential services and amenities, making them the cornerstone of modern urban planning, especially in densely populated cities.

#### 6.3.2.1 Mobility Hubs in Dense Cities

In dense urban areas, the challenges of traffic congestion, limited public space, and high demand for transportation options render mobility hubs particularly valuable (Arnold et al. 2023). Therefore, cities often struggle with overcrowded streets, inefficient transit systems, and lack of safe and accessible routes for pedestrians and cyclists. Mobility hubs offer a solution by integrating various transportation modes into a single accessible location, making it easier for residents to move around the city without relying on private cars. This integration not only alleviates congestion but also maximizes the use of limited urban spaces by consolidating the transportation infrastructure.

#### 6.3.2.2 Mobility Hubs in a 15-Minute City

The concept of a 15-minute city aims to create urban environments where all essential services, such as work, education, healthcare, and leisure, are accessible within a 15-minute walk or bike ride. Mobility hubs are instrumental in achieving this vision, particularly in dense cities, where spatial constraints require the efficient and creative use of infrastructure. By providing a network of interconnected hubs throughout the city, urban planners can ensure that all neighbourhoods are well-served by public transit and active transportation options, reducing the need for long commutes and minimizing the environmental impact of urban mobility.

#### 6.3.2.3 Challenges and Mitigation Strategies

Implementing mobility hubs as part of a 15-minute city strategy in dense urban areas presents unique challenges that require careful planning and policy intervention (Bencekri et al. 2024a). One of the primary challenges is the allocation of space to these hubs in crowded cityscapes. The study recommends that policymakers navigate

**Fig. 6.5** Mobility hubs design (Dilks 2021)

the complexities of land use, balancing the need for transportation infrastructure with other urban priorities, such as housing, commercial spaces, and green areas.

Cities can adopt several strategies to address these issues. First, integrating mobility hubs with existing infrastructure, such as underutilized parking lots or transit stations, can help maximize space efficiency (Bencekri et al. 2024a). For example, converting parking lots into mobility hubs that offer bike-sharing, electric vehicle charging, and connections to public transit can help repurpose valuable urban spaces for sustainable transportation. Additionally, leveraging smart city technologies can enhance the functionality of mobility hubs, providing real-time information on transit schedules, availability of shared bikes or scooters, and traffic conditions, thereby improving user experience and encouraging the use of multimodal transport.

### 6.3.3 AI-Based Allocation of Mobility Hubs

Bencekri et al. (2024a) presented a novel, three-tiered AI-based methodology for optimizing the location of carbon–neutral mobility hubs within a 15-minute city framework (Fig. 6.6). This approach integrates genetic algorithms (GAs) with ensemble-based suitability analysis and traffic assignment modelling to address several limitations of traditional site-selection methods. Typically, these methods focus on basic metrics, such as population density and travel time, often neglecting critical factors, such as carbon emissions and land use diversity.

The first stage of the adopted methodology involves the use of GAs to optimize travel times and transportation costs. GAs, which are well suited for solving complex problems with multiple variables, simulate natural selection processes to iteratively refine potential solutions. In this study, the GA was used to identify optimal hub locations by selecting sites that minimized travel time while maximizing network efficiency, thereby achieving a significant fitness value that underscores the effectiveness of this approach.

The second stage employed ensemble methods to conduct a suitability analysis, incorporating a range of factors weighted according to their contribution to carbon emissions. Ensemble methods such as Random Forest and Gradient Boosting enhance the predictive accuracy of the analysis by combining the strengths of multiple algorithms.

The final stage involved a quantitative post-assessment using traffic assignment modelling to evaluate the environmental and economic impacts of the selected hub locations.

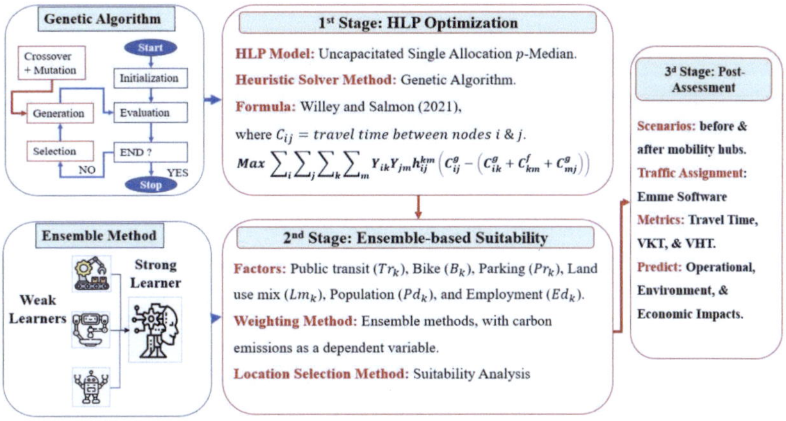


**Fig. 6.6** AI-based allocation of carbon–neutral mobility hubs (Bencekri et al. 2024b)

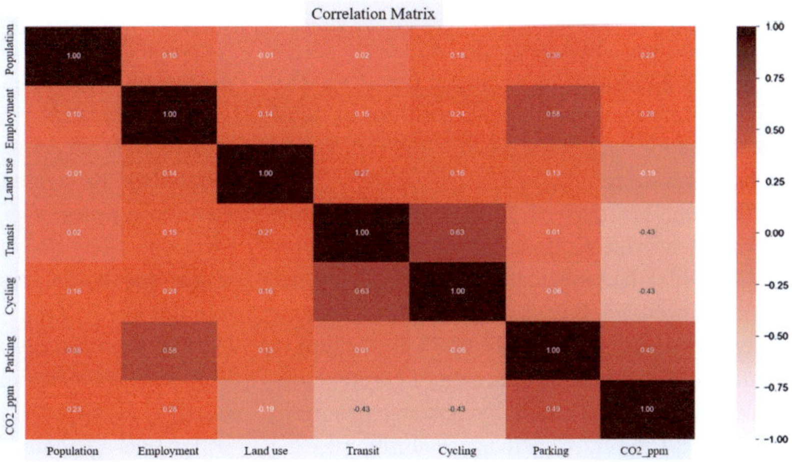


**Fig. 6.7** Correlation matrix (Bencekri et al. 2024b)

### 6.3.4 Results and Discussion

Using a correlation analysis (Fig. 6.7), the study found that carbon emissions were positively correlated with parking accessibility (correlation coefficient of 0.49), indicating that areas with higher parking accessibility tend to have higher carbon emissions. Conversely, there was a negative correlation between carbon emissions and public transit and cycling accessibility (correlation coefficients of -0.43 and -0.30, respectively), highlighting the importance of promoting active transportation and public transit as effective strategies for reducing carbon footprints. These findings underscore the importance of carefully considering these factors in the site selection process to enhance the environmental sustainability of urban mobility hubs.

In the first stage of the study, genetic algorithms (GAs) were utilized to identify optimal mobility hub sites by minimizing travel time and maximizing network efficiency. GA algorithm is known for their ability to solve complex optimization problems and to systematically refine a diverse array of candidate locations. The final selection focused on high-traffic areas, which offered the greatest potential for reducing transportation costs and improving network efficiency. The optimization process achieved a notable fitness value of 77,000,000, demonstrating the effectiveness of this approach in selecting highly efficient hub sites.

The second stage involved an ensemble-based suitability analysis, which assessed potential hub locations using a combination of advanced

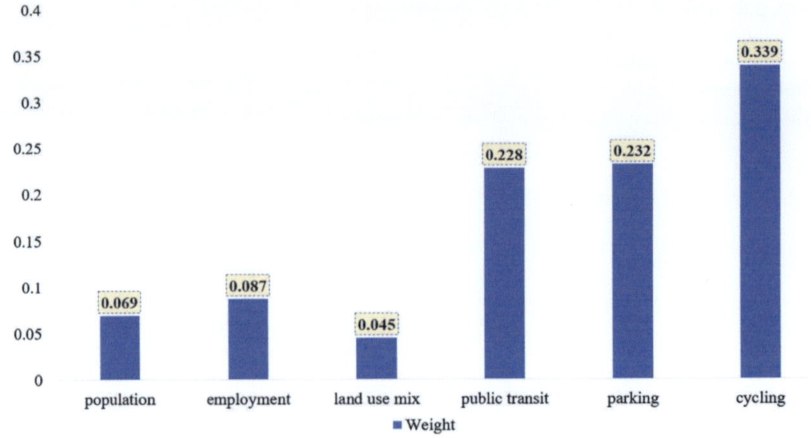


**Fig. 6.8** Ensemble algorithms derived weights

AI techniques, including the Random Forest Regressor, XGBoost, and Gradient Boosting. This study utilized these ensemble algorithms to calculate the importance of various factors influencing carbon emissions.

The weights derived from these algorithms (Fig. 6.8) provided a data-driven basis for determining which factors most significantly impacted carbon emissions. For example, cycling accessibility had the highest weight (0.3), indicating its strong influence on the reduction of carbon emissions. This was followed by parking proximity (0.232) and public transit accessibility (0.228), both of which also play critical roles in the sustainability of mobility hubs. Factors such as land-use mix, population, and employment densities had lower weights, suggesting that they were less influential than transportation-related factors.

These weighted factors were then applied to a suitability analysis, which generated scores to identify the most appropriate locations for mobility hubs (Fig. 6.9). The suitability analysis scores, which ranged from higher values for more suitable locations to lower values for less suitable locations, indicated the overall viability of each site based on its ability to meet the sustainability and efficiency criteria. The analysis identified several key areas within the study region that scored high, indicating their suitability for hosting carbon–neutral mobility hubs. For instance, areas with high cycling accessibility and well-connected public transit networks tended to receive the highest scores, making them prime hub placement candidates.

The final stage of the study involved a quantitative post-assessment using traffic assignment modelling to evaluate the environmental and economic impacts of the selected mobility hub locations. The findings, shown in Table 6.1, reveal substantial reductions in vehicle kilometres travelled (VKT) and vehicle hours travelled (VHT), with a daily reduction of 771,074 VKT. This reduction translated into significant annual savings of approximately 225.9 million USD, which was attributed to operational cost savings, travel time reductions, improved safety, and decreased air pollution. These results underscore the substantial economic and environmental benefits of strategically placed mobility hubs, highlighting their critical role in promoting sustainable urban mobility (Bencekri et al. 2024a).

### 6.3.5 Implications and Critics

The findings of Bencekri et al. (2024a) contribute significantly to the ongoing discourse on sustainable urban planning, particularly in the context of 15-minute cities. By prioritizing carbon emissions in the site selection process, this methodology aligns with broader environmental goals such as those outlined in the Paris Agreement. However, while this study offers valuable insights, it is essential to consider the practical

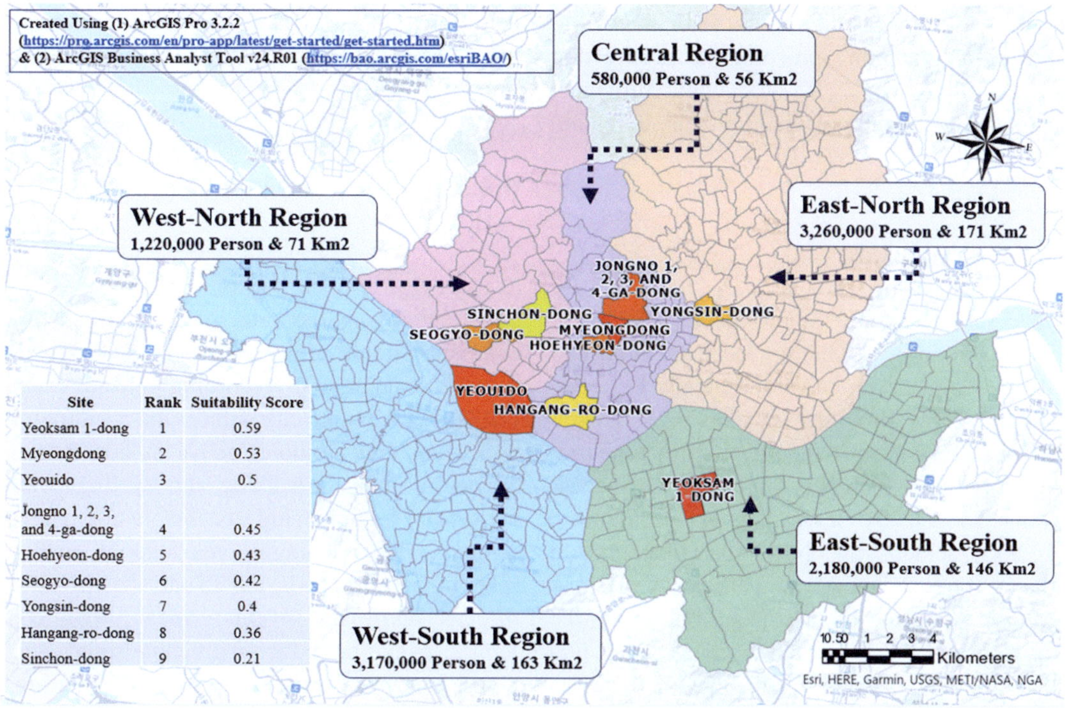


**Fig. 6.9** Final selected locations

**Table 6.1** Post-assessment analysis findings

| Mode | Number of trips (Before) | Number of trips (After) | Difference (Trips) | Difference (%) |
|---|---|---|---|---|
| Automobile | 21,196,831 | 21,146,175 | – 50,656 | – 37.75 |
| Taxi | 3,171,534 | 3,155,092 | – 16,442 | – 12.25 |
| Bus | 6,995,355 | 7,014,433 | + 19,078 | + 14.22 |
| Subway | 5,916,110 | 5,949,049 | + 32,939 | + 24.55 |
| Combined subway and bus | 2,833,118 | 2,848,199 | + 15,081 | + 11.24 |
| Total modal shift | | | 67,098 | |

challenges and potential limitations of implementing such AI-driven approaches in diverse urban contexts.

One critical perspective is the potential for inequitable outcomes if mobility hubs are inaccessible to all urban residents. As Tran and Draeger (2021) suggested, the success of the 15-minute city model depends on its ability to serve diverse populations, including those in underserved or marginalized communities. Moreover, retrofitting existing urban environments to fit the 15-minute city model presents significant challenges, particularly in cities with entrenched car-dependent infrastructure. These challenges underscore the need for context-sensitive strategies to account for the unique characteristics of each city.

Additionally, although AI techniques offer powerful tools for optimizing urban planning, they are not without limitations. Reliance on data-driven models may overlook the qualitative aspects of urban life, such as cultural and historical contexts, which are crucial in creating vibrant and liveable communities. Future research should explore how AI can be integrated with participatory planning processes to ensure that mobility hubs are designed with inputs from local communities, thereby enhancing their social and environmental relevance.

### 6.3.6 Conclusion

The 15-minute city offers a promising pathway for decarbonization by promoting localized living and reducing reliance on private vehicles. Bencekri et al. (2024a) provided a robust framework for optimizing the placement of carbon–neutral mobility hubs, which is essential for the success of this model. By leveraging AI techniques, this study demonstrates how urban planners can enhance the sustainability and efficiency of transportation systems, thereby contributing to the broader goals of climate change mitigation and urban resilience. However, the practical implementation of these strategies requires careful consideration of social equity and the unique challenges of retrofitting the existing urban environments. As cities continue to adapt to the realities of climate change, the principles and methodologies outlined in this study will be critical in shaping the future of urban mobility and sustainability.

## 6.4 Walkability Planning Support System

### 6.4.1 Introduction

The 15-minute city is an innovative urban planning model designed to create compact, self-sufficient neighbourhoods. Central to the success of the 15-minute city is the concept of walkability, the ease with which residents can move around their neighbourhoods on foot. Walkability not only reduces dependence on cars, but also fosters social interaction, supports local businesses, and contributes to a healthier urban environment. This chapter explores the development and application of a Walkability Planning Support System (W-PSS) as a key tool in realizing the 15-minute city, drawing primarily on the study by Bencekri et al. (2024b), and supplemented by other relevant academic sources.

### 6.4.2 Walkability in a 15-Minute City

Walkability is a cornerstone of the 15-minute city, embodying the principle that urban environments should prioritize pedestrian movement over vehicular traffic. According to Moreno et al. (2021), a 15-minute city reduces the need for long commutes and promotes active modes of transportation, such as walking and cycling, which are crucial for achieving sustainable urban development. By encouraging residents to walk to meet their daily needs, the 15-minute city model directly addresses several critical urban challenges, including reducing greenhouse gas emissions, improving public health, and fostering a sense of community.

However, achieving high levels of walkability requires more than well-designed sidewalks and crosswalks. This necessitates a comprehensive approach that integrates various urban metrics, socioeconomic factors, and the subjective experiences of pedestrians (Venerandi et al. 2024). Traditional walkability measures, such as the Walk Score, often fall short in capturing the complex interplay of these factors, leading to a need for more sophisticated tools, such as the Walkability Planning Support System (Bencekri et al. 2024c).

### 6.4.3 Methodology for the W-PSS

Bencekri et al. (2024b) introduced a pioneering W-PSS, designed to enhance walkability in urban areas. This system is particularly relevant to the 15-minute city concept as it provides urban planners with the tools needed to assess and improve walkability at a granular level. The W-PSS framework is built around a newly developed Walkability Index (WI) that integrates both objective and subjective measures of walkability. This index

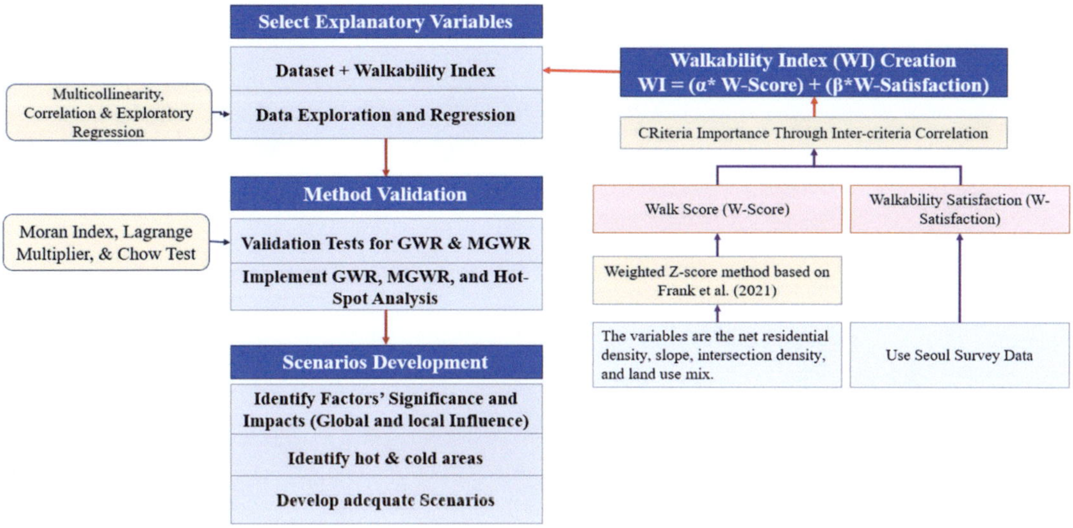


**Fig. 6.10** WPSS methodology (Bencekri et al. 2024c)

includes a Walk Score (W-score) based on key urban metrics, such as intersection density, land-use mix, and residential density, as well as a Walk Satisfaction (W-satisfaction) score, which reflects pedestrian perceptions collected through surveys.

To ensure that the W-PSS provides an accurate and balanced assessment of walkability, this study employs the CRiteria Importance Through Inter-criteria Correlation (CRITIC) method. This method allows for the assignment of objective weights to the various factors influencing walkability (Krishnan 2024), ensuring that the final Walkability Index reflects the unique characteristics of different urban contexts. Figure 6.10 illustrates how these components are combined to create a comprehensive measure of walkability, which is both data-driven and responsive to the lived experiences of the pedestrians.

W-PSS further refines its analysis using advanced spatial tools such as Geographic Weighted Regression (GWR) and Multiscale GWR (MGWR). These models allow for the examination of spatial variations in walkability, providing insights into how different factors influence pedestrian movement in various parts of a city (Bencekri et al. 2024c). For instance, in this study, the MGWR model, which demonstrated an adjusted $R^2$ of 67.3%, outperformed GWR in capturing the complex multiscale relationships between walkability and its determinants. These tools enable urban planners to identify areas where walkability can be improved through targeted interventions.

### 6.4.4 W-PSS for a 15-Minute Seoul

The W-PSS was applied in Seoul to demonstrate its practical utility in enhancing walkability within a 15-minute city framework. This study, conducted by Bencekri et al. (2024b), provided several key insights into how various factors influence walkability in different urban contexts.

#### 6.4.4.1 Population, Employment, and Retail

The analysis revealed that increasing population and employment densities generally enhance walkability, particularly in urban centres, where these densities support a vibrant pedestrian environment (Bencekri et al. 2024c). However, the study also cautioned against over-densification, which could lead to congestion and reduced walkability. Commercial density had a more complex relationship with walkability. Although they often support pedestrian activities by providing destinations, they also contribute to congestion in certain areas (Figs. 6.11 and 6.12). This suggests the need

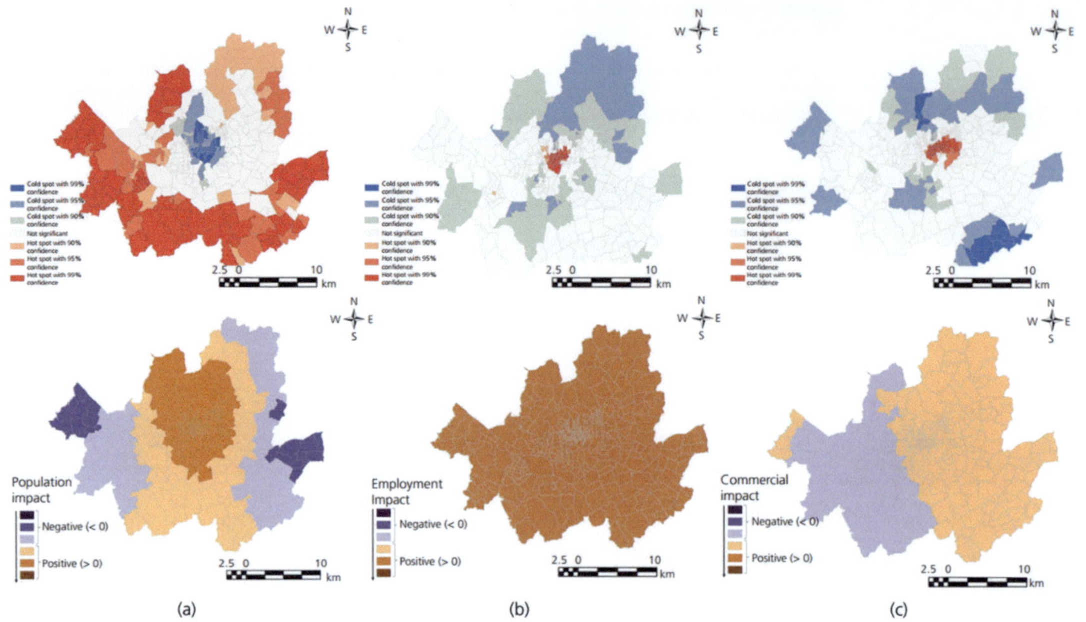


**Fig. 6.11** Hotspot analysis and MGWR unscaled residuals: **a** population density analysis, **b** employment density analysis

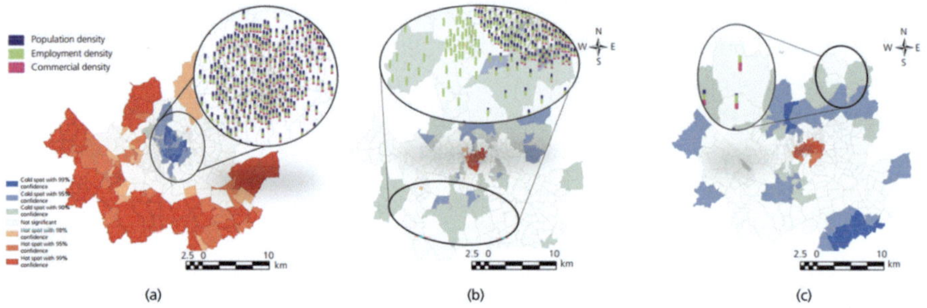


**Fig. 6.12** Density-basis scenario (population, employment, and commercial)

for balanced density management to optimize walkability.

#### 6.4.4.2 Land-Use Mix and Green Space

A well-balanced land-use mix is essential for fostering walkability, as it ensures that residents have easy access to a variety of services and amenities (Gehrke and Clifton 2019). Bencekri et al. (2024b) found that areas with a diverse mix of residential, commercial, and recreational spaces tended to have higher walkability scores. Additionally, the findings prove that the presence of green spaces significantly enhances the walking experience, providing both environmental and psychological benefits to pedestrians (Fig. 6.13). This finding supports the integration of green infrastructure into urban planning as a key component of a 15-minute city.

#### 6.4.4.3 Public Transit and Safety

The integration of public transit with pedestrian pathways has been identified as a critical factor for enhancing walkability (Wey and Chiu 2013). Areas with well-connected transit networks, particularly in the central business district (CBD), exhibited higher walkability scores. However, the results highlighted potential conflicts between cyclists and pedestrians (Bencekri et al. 2024c), particularly in areas where bike lanes were

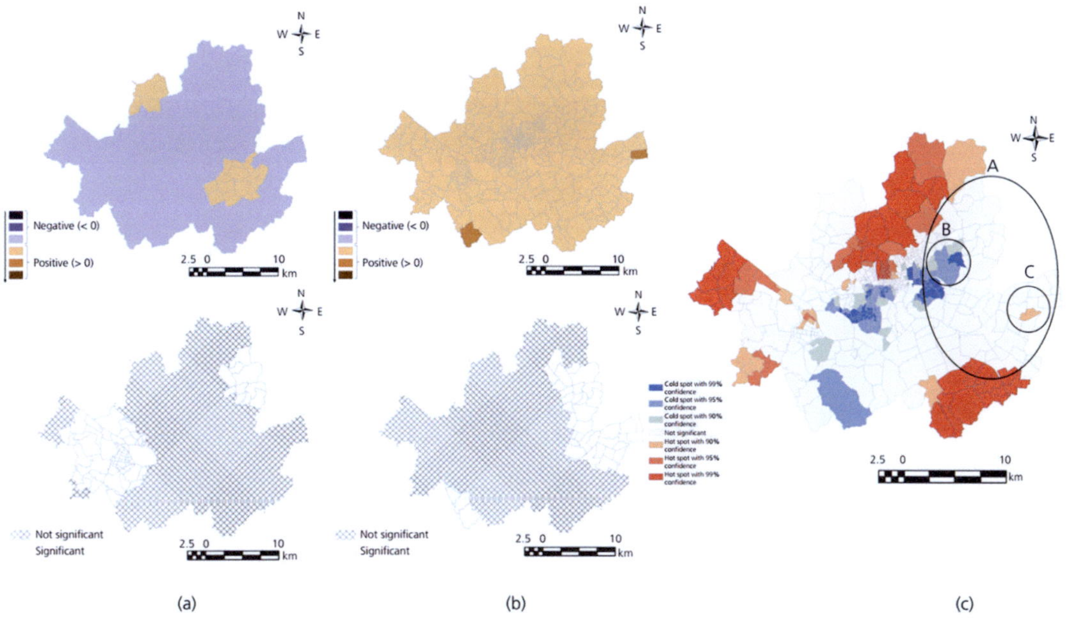


**Fig. 6.13** Land-use mix and greenness analysis: **a** land-use mix impact, **b** greenness index impact and **c** hotspot analysis of green areas

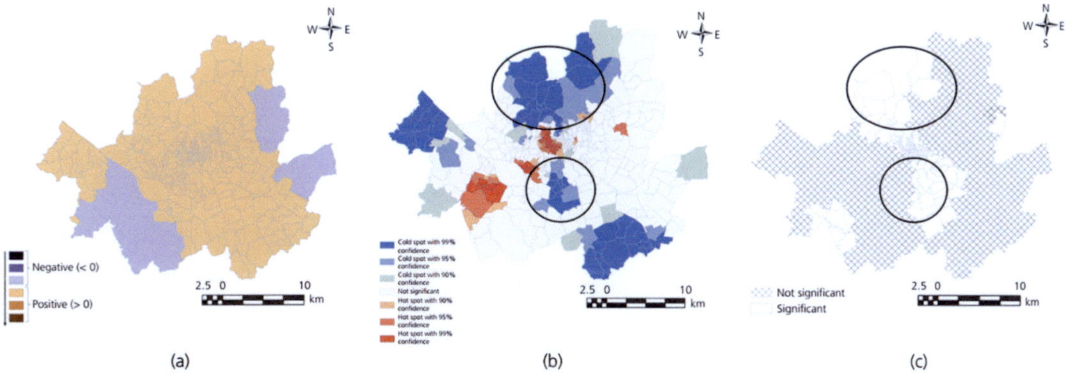


**Fig. 6.14** Public transit coverage analysis: **a** impact analysis, **b** hotspot analysis and **c** significance analysis of greenness

not well integrated with pedestrian pathways (Figs. 6.14 and 6.15). This finding underscores the need for careful planning to ensure that different modes of active transportation complement each other rather than compete.

#### 6.4.4.4 Cycling and Pedestrian Interaction

A significant finding of this study was the negative correlation between cycling infrastructure and walkability in some areas (Bencekri et al. 2024c). This suggests that poorly planned or overly congested cycling paths can detract from pedestrian experience, leading to conflicts between cyclists and walkers. To address this issue, the study recommends the development of a separate but interconnected infrastructure for cyclists and pedestrians, ensuring that both groups can coexist safely and comfortably (Figs. 6.15 and 6.16).

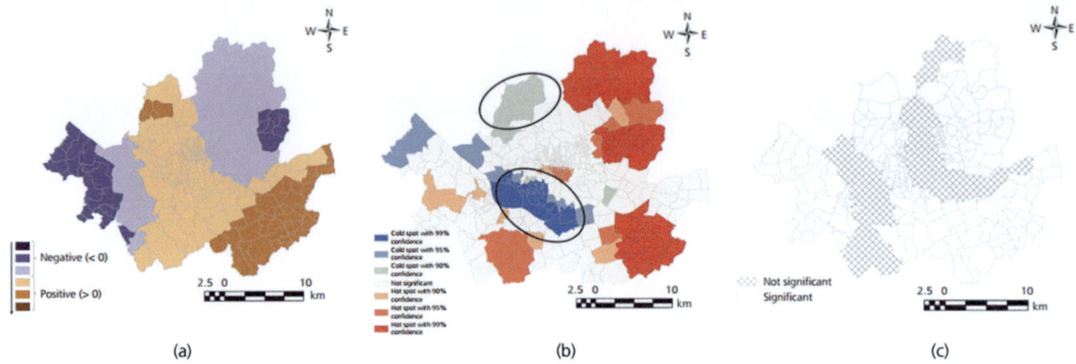


**Fig. 6.15** Road safety analysis: **a** impact analysis, **b** hotspot analysis and **c** significance analysis

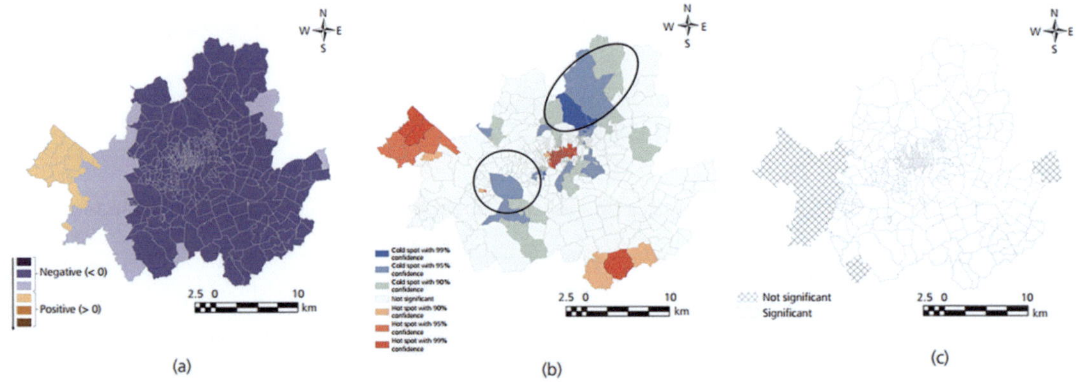


**Fig. 6.16** Shared bike analysis: **a** impact analysis, **b** hotspot analysis and **c** significance analysis

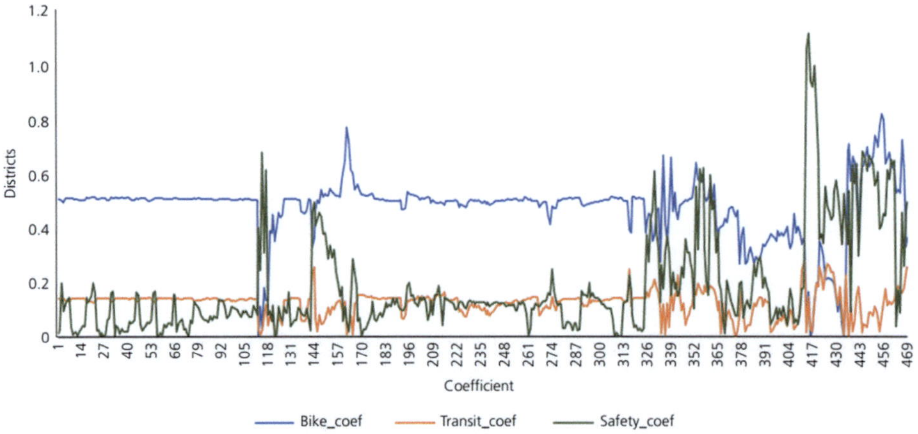


**Fig. 6.17** Comparison of coefficients' absolute values

## 6.4.5 Results and Implications

Bencekri et al. (2024b) identified safety as a critical factor that influences walkability. Figure 6.17 shows that safety concerns, particularly in areas with inadequate pedestrian infrastructure, significantly reduced walkability. Improvements in road safety, such as traffic calming measures, improved pedestrian crossings, and enhanced street lighting,

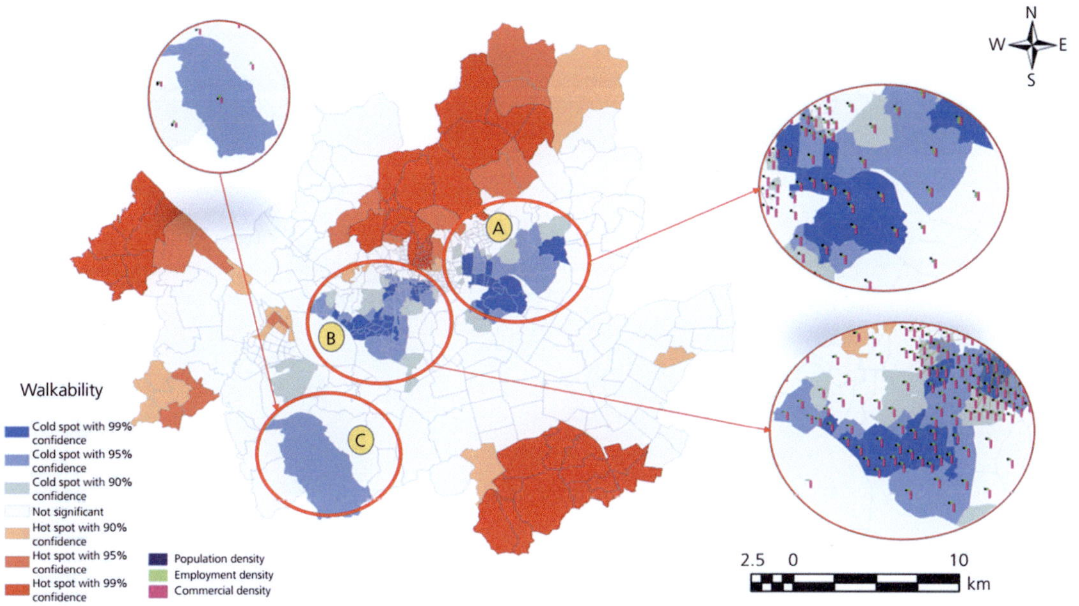


**Fig. 6.18** Transport-based scenario

are recommended to create safer environments for pedestrians (Kim et al. 2024). Public transit coverage, particularly in the CBD and peripheral areas, was found to have a positive impact on walkability.

Figure 6.18 shows that well-integrated public transit systems can enhance walkability by providing convenient and accessible travel options. However, the study also found that, in some regions, such as Region C, public transit had a negative impact on walkability, suggesting a need for further assessment of transit accessibility and convenience.

Cycling infrastructure also plays a complex role in walkability. As Fig. 6.17 illustrates, there is often conflict between cyclists and pedestrians, particularly in areas where bike lanes encroach on pedestrian spaces. This suggests the need for a better-designed segregated infrastructure that can accommodate both modes of transport without compromising pedestrian safety.

In addition, Fig. 6.19 shows the relationship between walkability and public transit frequency. The analysis revealed that in regions such as Jongno-gu and Jung-gu, high walkability scores correlated with relatively low public transit frequencies, indicating that further enhancements in public transit could lead to even higher levels of walkability. This finding suggests that increasing the frequency and coverage of public transit services can significantly boost pedestrian activity, particularly in areas where public transit is underutilized.

These findings from the W-PSS application in Seoul have important implications for urban planning and policy, particularly within the context of the 15-minute city (Bencekri et al. 2024c). By providing a detailed and context-sensitive assessment of walkability, the W-PSS offers a powerful tool for urban planners aiming to create pedestrian-friendly environments. The ability to simulate different planning scenarios using the W-PSS allows flexible and adaptive strategies that can be tailored to the specific needs of different urban areas.

One of the key strengths of the W-PSS is its ability to integrate both objective and subjective experiences, thus providing a holistic view of walkability. This dual approach ensures that urban planning decisions are based not only on physical metrics, but also on the lived experiences of resi-

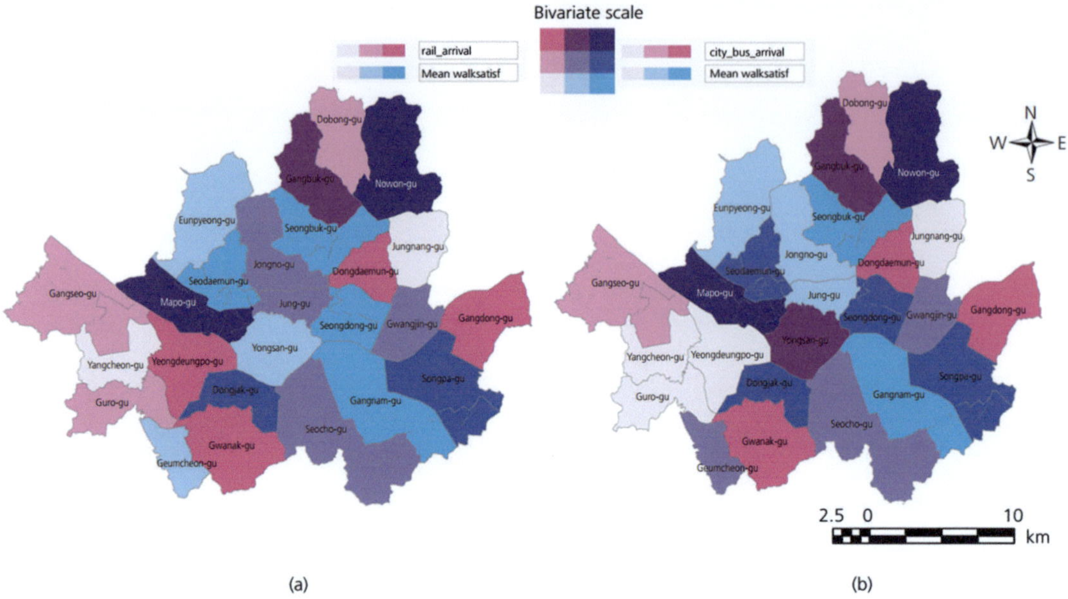


**Fig. 6.19** Walkability and public transit frequency (number of departs per day): **a** walkability and rail frequency and **b** walkability and bus frequency

dents. As cities around the world seek to reduce their carbon footprints and improve the quality of life of their residents, tools such as the W-PSS will be essential in guiding the way forward.

Moreover, W-PSS highlights the importance of considering the interplay between different modes of transportation. In the context of a 15-minute city, it is crucial to ensure that walking, cycling, and public transit are seamlessly integrated to create a cohesive and accessible urban environment. The findings from this study suggest that achieving this integration requires careful planning and continuous monitoring, with a focus on balancing the needs of different users.

### 6.4.6 Conclusion

The Walkability Planning Support System (W-PSS) developed by Bencekri et al. (2024b) represents a significant advancement in the field of urban planning, particularly within the framework of the 15-minute city. By providing a detailed and context-sensitive assessment of walkability, the W-PSS supports the goals of the 15-minute city, promoting sustainable, inclusive, and vibrant urban environments. The findings from this study underscore the importance of tailored data-driven strategies in improving walkability, highlighting the need for integrated planning that considers both the physical and experiential aspects of pedestrian environments. As cities continue to evolve, tools such as the W-PSS will be essential for creating sustainable, liveable urban spaces that prioritize the needs of pedestrians.

## 6.5 Bikeability Planning Support System

### 6.5.1 Introduction

Another central pillar for the 15-minute city is the promotion of sustainable transportation modes, including cycling, which reduces reliance on automobiles and fosters healthier, more connected communities (Martin 2024). This chapter explores how a Planning Support System (PSS) designed to enhance bikeability can serve as a critical tool in realizing a 15-minute city. Drawing on the study by Bencekri et al. (2023b), which developed a PSS to boost bikeability using Seoul as a use case, this

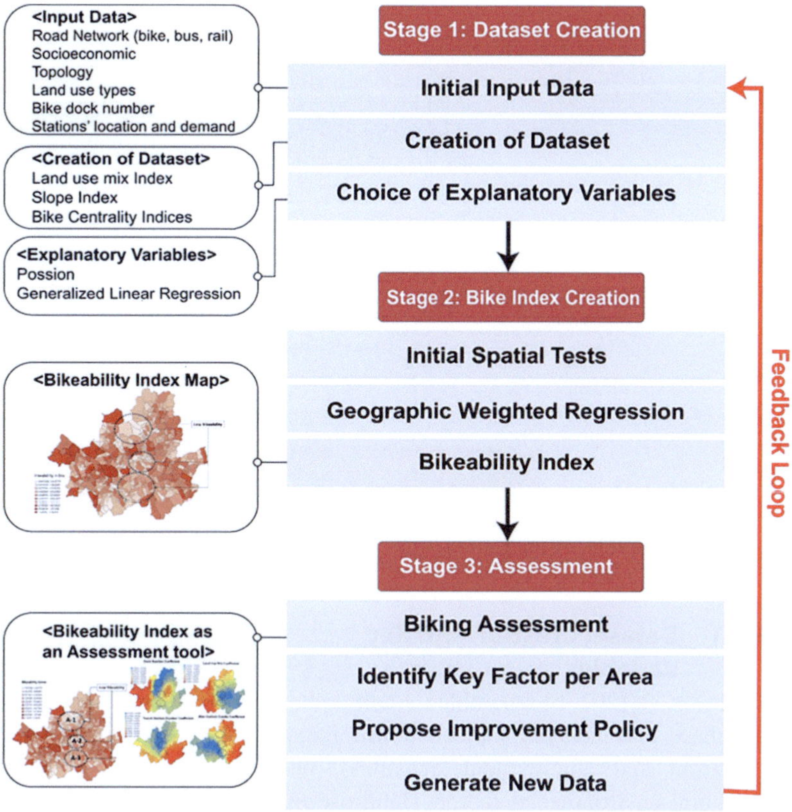


**Fig. 6.20** Bike planning support system flowchart

chapter will examine the methodologies, findings, and implications of this system, linking them to the broader goals of the 15-minute city.

### 6.5.2 Bikeability in the 15-Minute City

Bikeability is a crucial aspect of a 15-minute city, emphasizing the need for urban environments to support cycling as a viable and attractive mode of transportation (Bencekri et al. 2023b). According to the principles of the 15-minute city, cycling should not only be safe and convenient, but also integrated with other forms of transportation, such as public transit, to facilitate seamless mobility. As cities worldwide strive to reduce carbon emissions, alleviate traffic congestion, and improve public health, enhancing non-motorized means of mobility, such as bikeability, has become a key strategy in achieving these goals (Mishra et al. 2024).

Bencekri et al. (2023b) highlight the importance of bikeability by demonstrating how a well-designed PSS can improve the effectiveness and performance of a shared-bike system in an urban setting. In Seoul, the adoption of such a system aligns with the city's broader objectives of mitigating traffic congestion, reducing air pollution, and enhancing the quality of life for its residents.

### 6.5.3 Method of the Bikeability Planning Support System

The Bikeability Planning Support System (PSS) developed by Bencekri et al. (2023b) was designed to enhance the cycling environment in Seoul by addressing key factors that influence bikeability. The PSS was structured around four main components: creating a comprehensive dataset, building a bikeability index, assessing the shared-cycling system, and implementing various improvement scenarios (Fig. 6.20).

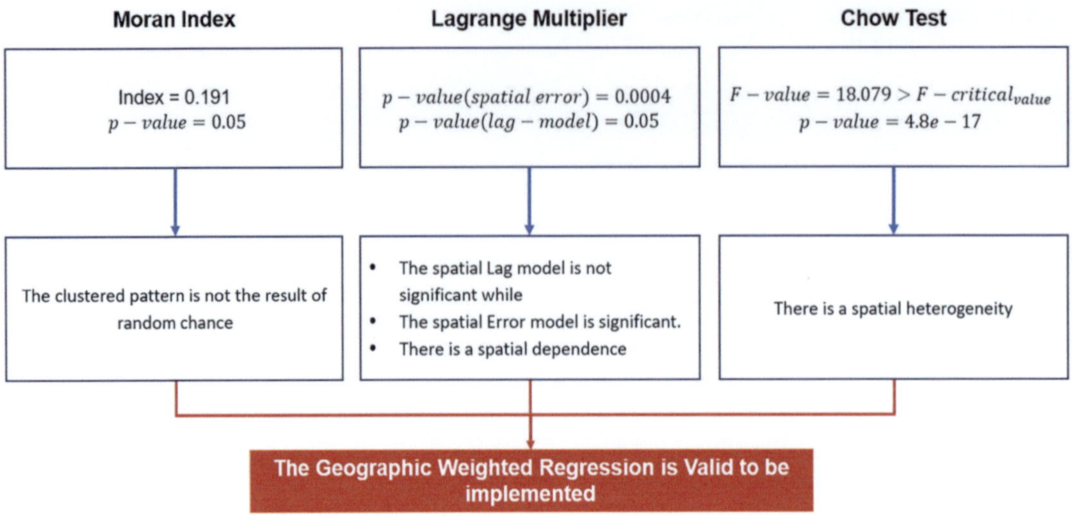


**Fig. 6.21** Development of the GWR model

#### 6.5.3.1 Dataset Creation and Key Variables

Bencekri et al. (2023b) assembled a comprehensive dataset that included socioeconomic factors (population and employment densities), built and natural environmental factors (land-use mix and slope), public transit and cycling infrastructure (station locations, number of docks, and bike-transit distances), and economic activity (retail consumption). These factors were chosen based on their significant influence on cycling behaviour and accessibility (Bencekri et al. 2023b).

#### 6.5.3.2 Developing the Bikeability Index

A central element of the PSS is the development of a bikeability index that quantifies the ease and convenience of cycling in the different districts of Seoul. This index was constructed using the GWR model (Fig. 6.21), which accounts for spatial variability in the factors affecting bikeability.

The GWR model showed high accuracy, as evidenced by a high R-squared value (83.2%) and a lower Akaike Information Criterion (AICc) compared to the traditional Poisson models (Table 6.2, Figs. 6.22 and 6.23). The final bikeability index incorporates variables, such as employment and population densities, land-use mix, slope, retail consumption, and centrality indices, reflecting the accessibility of bike stations (Bencekri et al. 2023b).

**Table 6.2** Results of the two models (GWR and Poisson)

| Evaluation factor | GWR model | Poisson model |
|---|---|---|
| *R*-square (%) | 83.2 | 82.7 |
| AICc | 4,465.41 | 8,931.35 |

#### 6.5.3.3 Scenario Development and Assessment

The PSS also included a scenario analysis component (Fig. 6.24), and different strategies for improving bikeability were simulated and evaluated (Bencekri et al. 2023b). Three scenarios were developed: increasing bike-only lanes, increasing both bike lanes and docking stations while decreasing the distance to transit stations, and a combination of adding docks, increasing bike lanes, and reducing transit distances (Fig. 6.25).

The second scenario, which focused on increasing bike lanes and docking stations while enhancing connectivity with transit, was found to be the most effective (Fig. 6.26). It improves all key centrality measures (betweenness, closeness, and gravity), as shown in Fig. 6.27, and significantly boosts the bikeability index across cities (Fig. 6.28).

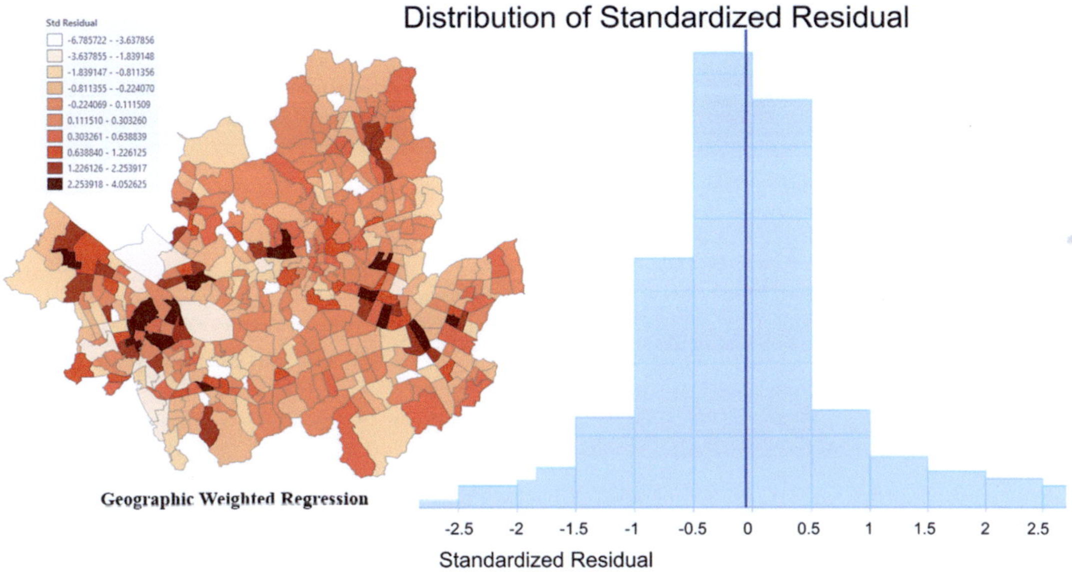


**Fig. 6.22** GWR standard residual map

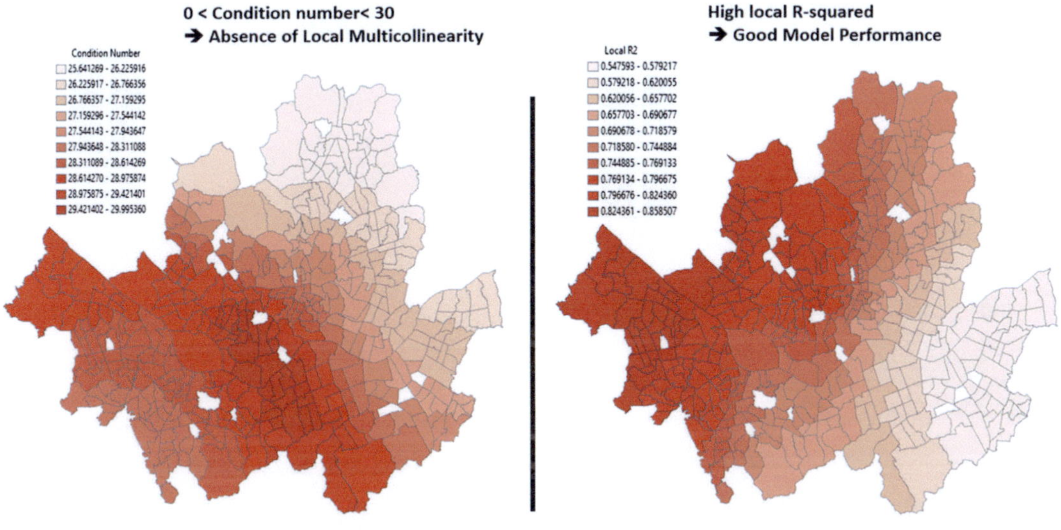


**Fig. 6.23** GWR model's performance

#### 6.5.3.4 Spatial Analysis and Validation

Spatial analysis plays a crucial role in validating the effectiveness of the proposed scenarios. Bencekri et al. (2023b) conducted spatial validation tests, including tests for spatial autocorrelation, dependence, and heterogeneity, which confirmed the appropriateness of the GWR model for analysing bikeability in Seoul (Fig. 6.21). Moran's I, Lagrange Multiplier, and Chow tests all indicated significant spatial effects, justifying the use of spatially weighted models in the analysis.

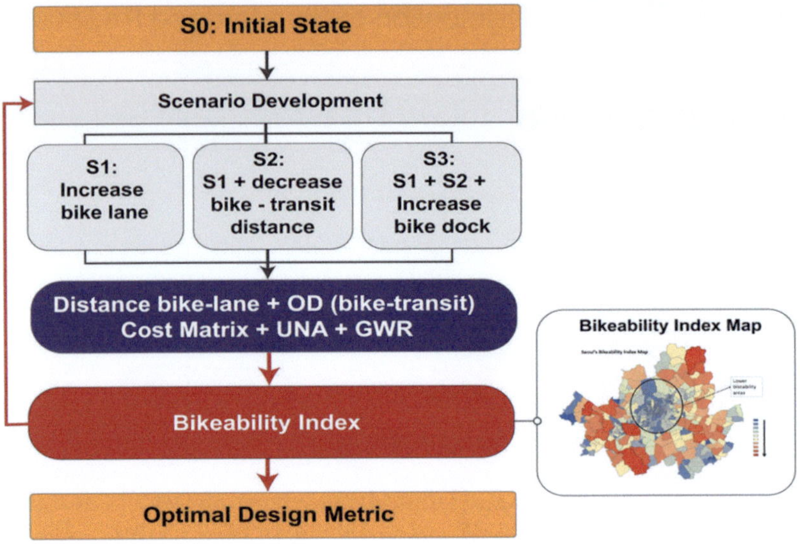


**Fig. 6.24** Scenario-based analysis

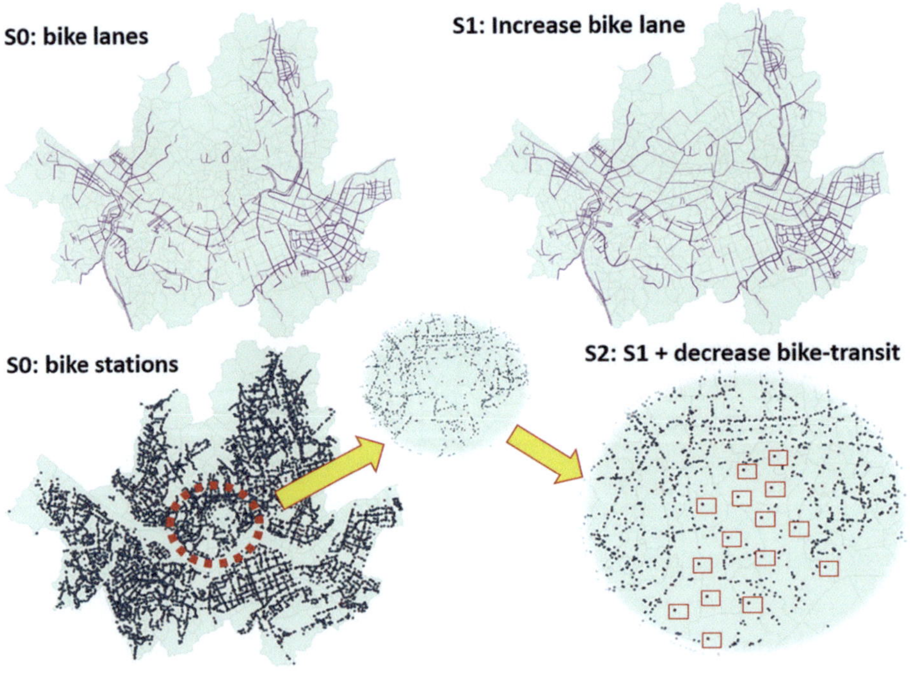


**Fig. 6.25** Bike-developed scenarios for the PSS

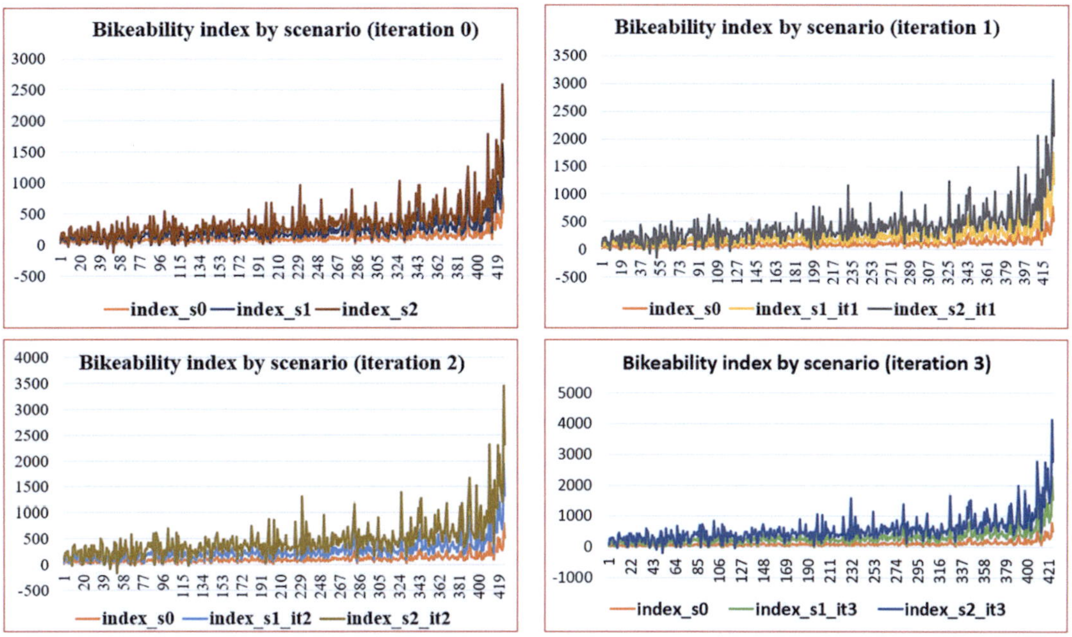


**Fig. 6.26** Bikeability index by scenario and iteration

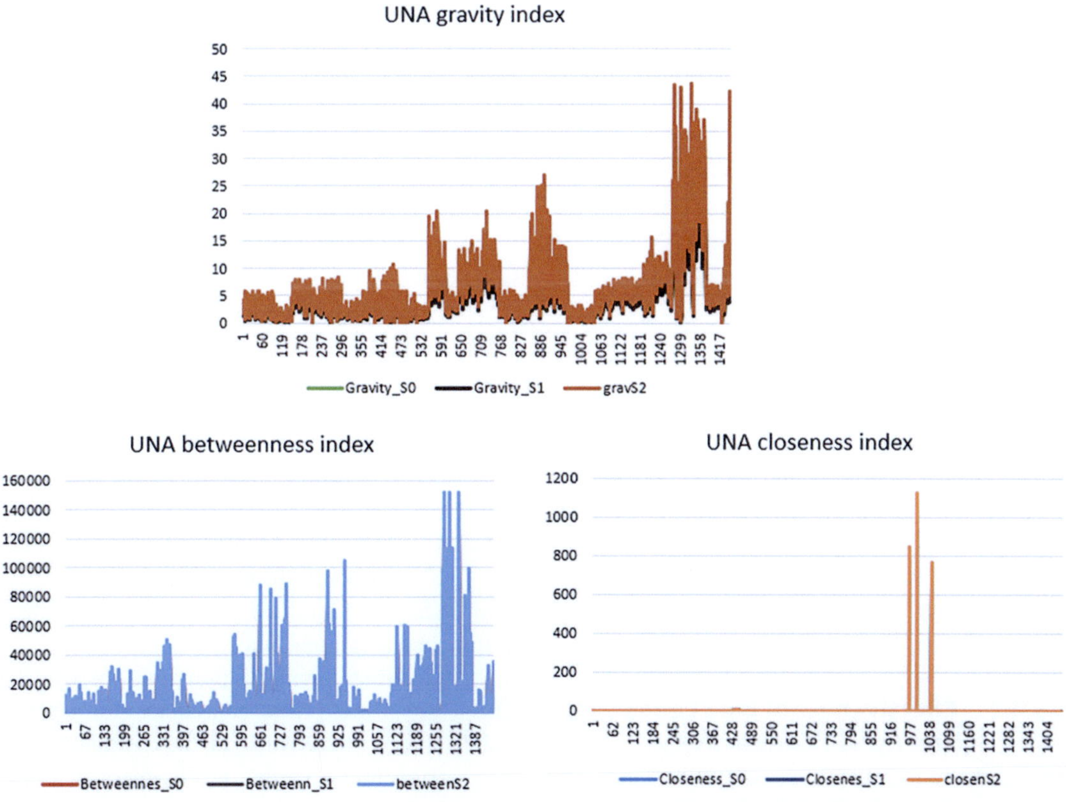


**Fig. 6.27** Scenarios' centrality indices results using the UNA tool

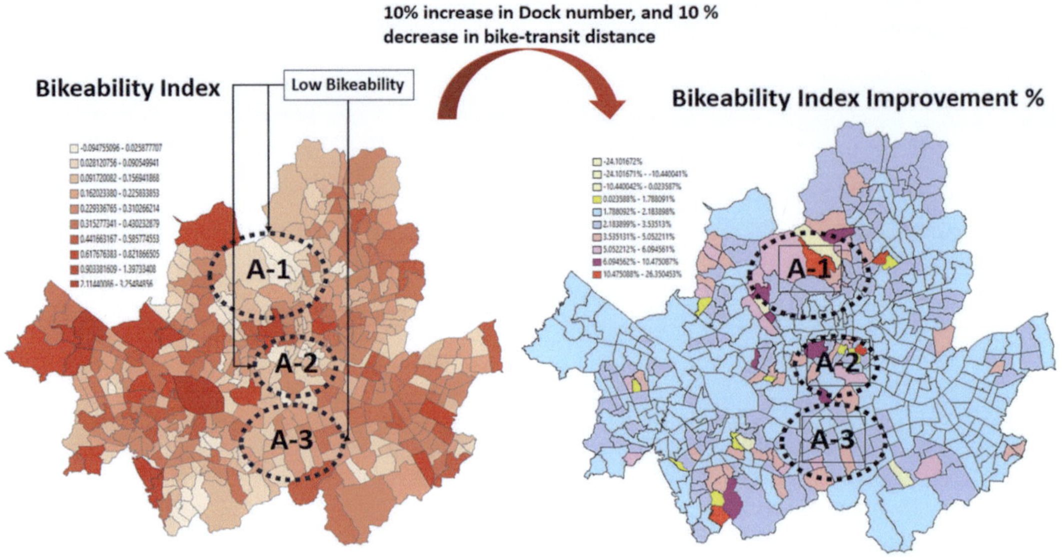


**Fig. 6.28** Results of the scenario 2

### 6.5.4 Findings and Implications of Bikeability PSS for Seoul

The findings of this study have significant implications for the implementation of a 15-minute city (Bencekri et al. 2023b). By providing a detailed assessment of bikeability and identifying the most effective interventions for enhancing cycling infrastructure, PSS offers a powerful tool for urban planners. Key findings include.

#### 6.5.4.1 Impact of Centrality and Accessibility

The analysis revealed that centrality measures (betweenness, closeness, and gravity) are crucial determinants of bikeability. Improving the centrality of bike stations by optimizing their locations within the urban network can significantly enhance the overall accessibility of cycling infrastructure (Fig. 6.29).

#### 6.5.4.2 Effectiveness of Bike Lanes and Transit Connectivity

The study found that increasing bike lanes and improving the connectivity between bike stations and public transit hubs are among the most effective strategies for boosting bikeability. These interventions not only make cycling more convenient but also support the integration of cycling with other forms of transportation, a key principle of the 15-minute city (Fig. 6.29).

#### 6.5.4.3 Spatial Heterogeneity in Bikeability

GWR analysis highlighted the significant spatial heterogeneity in the factors influencing bikeability across different districts of Seoul. This suggests that a one-size-fits-all approach to cycling infrastructure may not be effective; instead, tailored strategies that account for local conditions are necessary to optimize bikeability (Fig. 6.29).

#### 6.5.4.4 Potential for Policy Interventions

The PSS also identified areas where the bikeability index is high, but actual bike usage was low, indicating untapped potential for cycling. These areas are prime candidates for targeted policy interventions such as the introduction of cycling incentives or the development of exclusive bike lanes.

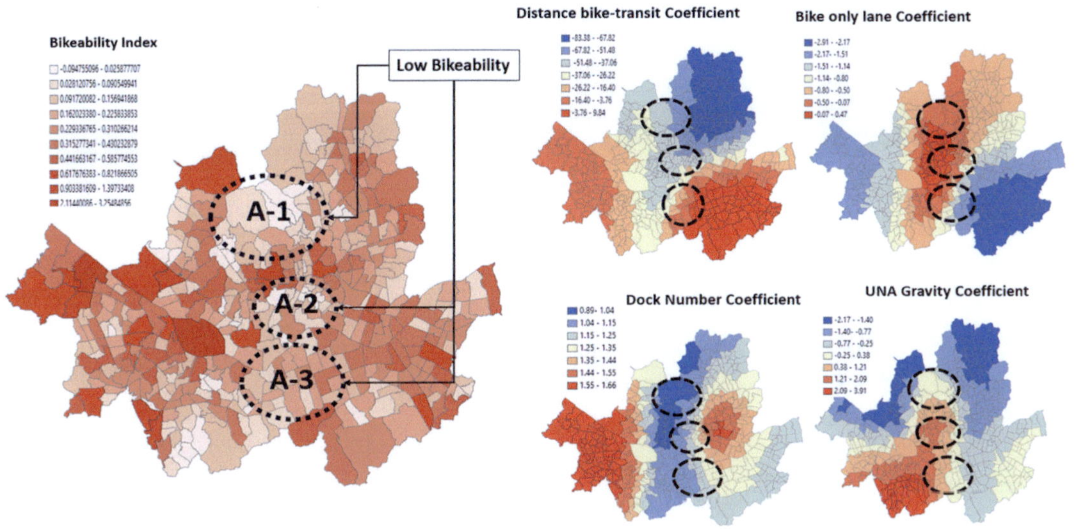


**Fig. 6.29** Assessment tools for bikeability index

### 6.5.5 Conclusion

The Bikeability Planning Support System (PSS) developed by Bencekri et al. (2023b) represents a significant advancement in urban planning, particularly within the context of a 15-minute city. By providing a comprehensive, data-driven assessment of bikeability and offering actionable insights for improving cycling infrastructure, the PSS supports the broader goal of creating sustainable, liveable urban environments (Bencekri et al. 2023b). The findings from this study underscore the importance of integrating cycling with other forms of transportation, enhancing the accessibility of bike stations, and tailoring interventions to local conditions. As cities continue to evolve and strive towards sustainability, tools such as PSS will be essential in guiding the development of bike-friendly, 15-minute urban spaces.

## 6.6 15-Minute City for Economic Gain

### 6.6.1 Introduction

The 15-minute city model aims to create sustainable, resilient, and vibrant urban communities by fostering a stronger connection between residents and their neighbourhoods (Zong 2023). While the primary focus of a 15-minute city has often been on sustainability and quality of life, its potential for economic gain is equally significant. The strategic placement of bike stations and mobility hubs is crucial in realizing these economic benefits, as these infrastructures serve as the linchpins of a multimodal transportation network that can stimulate local economies, enhance connectivity, and attract investments. Based on the study by Bencekri et al. (2024c), this chapter explores how optimizing the locations of bike stations and mobility hubs can generate substantial economic gains, drawing on recent research that employs advanced spatial and economic modelling techniques.

### 6.6.2 The Economic Role of Cycling in the 15-Minute City

Bike-sharing systems are an essential component of a 15-minute city, particularly in densely populated urban areas where space is limited, and transportation demands are high. By providing residents with convenient access to bicycles, these systems encourage cycling as a viable alternative to driving (Bencekri et al. 2024d). This

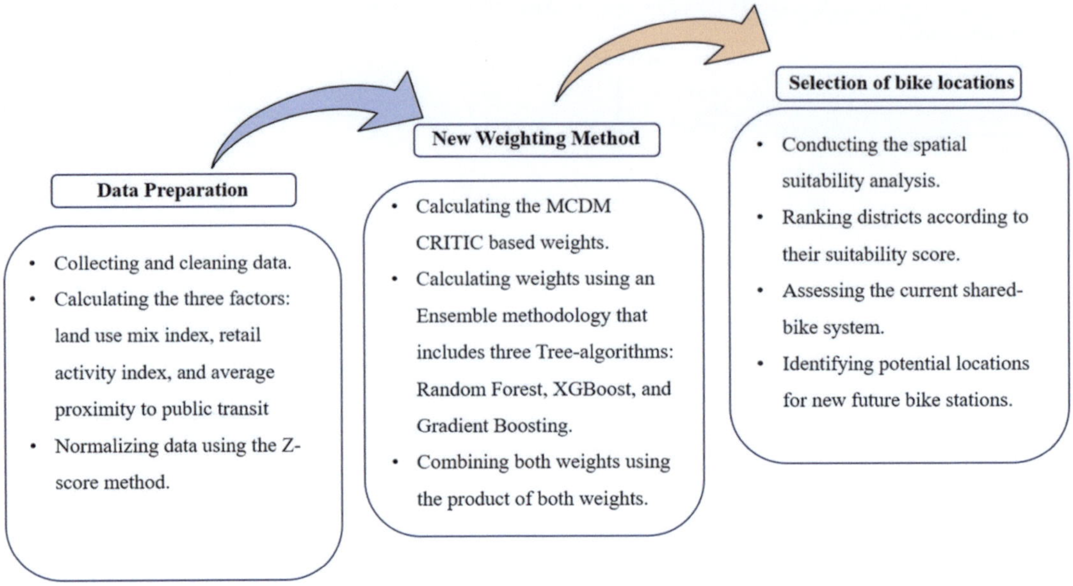


**Fig. 6.30** Methodology of the allocation of bike station for economic gain

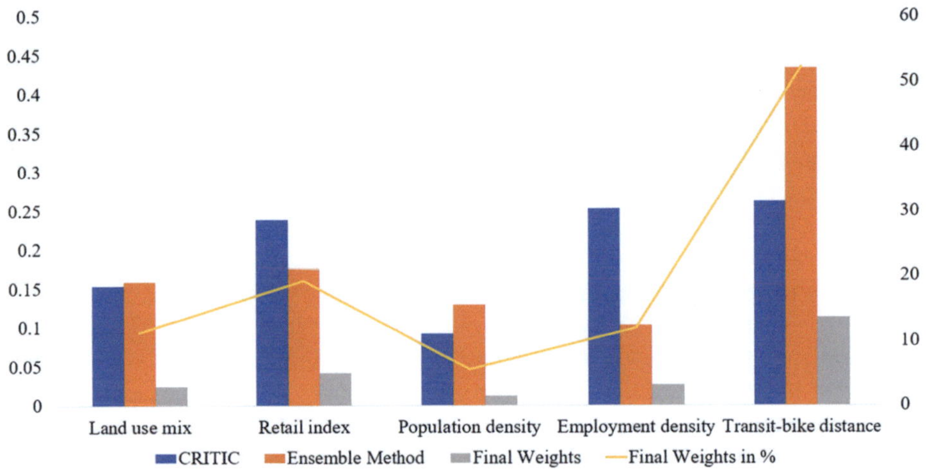


**Fig. 6.31** Computed weights for considered factors

shift not only helps to reduce traffic congestion and lower carbon emissions but also stimulates economic activity by increasing foot traffic in areas surrounding bike stations.

The study by Bencekri et al. (2024c) offers a detailed analysis of how optimizing the locations of bike stations can maximize economic benefits within a 15-minute city framework. The research methodology (Fig. 6.30) highlights the importance of incorporating various urban factors such as land-use mix, retail activity, transit proximity, and population and employment densities into the decision-making process for bike station placement. By applying a novel weighting method that combines the CRiteria Importance Through Inter-criteria Correlation (CRITIC) method (Fig. 6.31) with ensemble-based predictive algorithms (such as Random Forest, XGBoost, and Gradient Boosting), this study assessed the economic potential of different locations for bike stations with a high degree of precision.

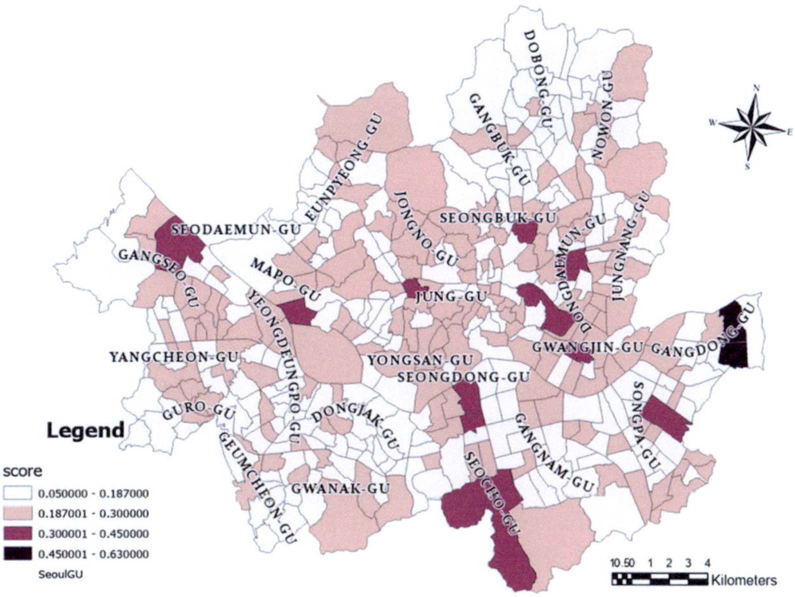

Fig. 6.32 Suitability score results

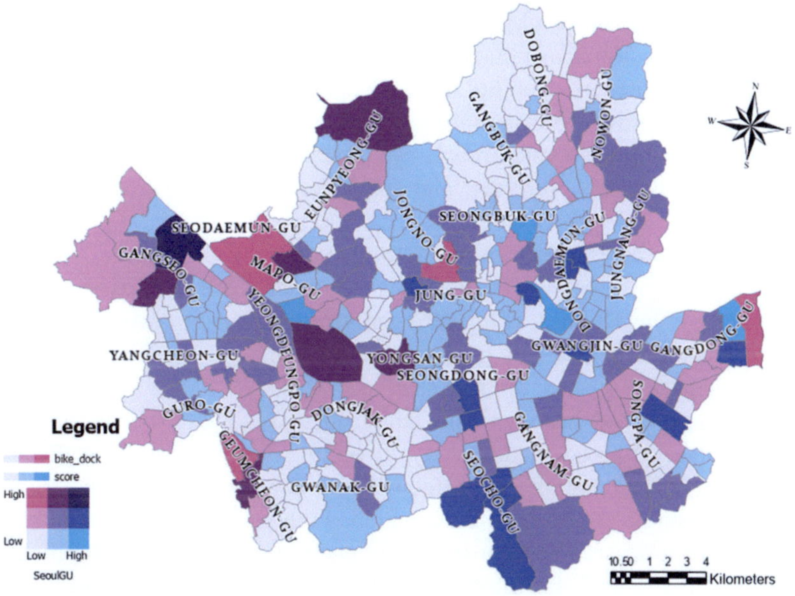

Fig. 6.33 Bivariate representation of bike docks and the suitability scores

The suitability scores generated by the study revealed considerable variability across different districts, underscoring the importance of a tailored approach to bike station placement (Fig. 6.32 and Fig. 6.33). In areas where the land-use mix is less balanced or retail activity is lower, the economic benefits of bike stations may be less pronounced. However, even in these areas, strategically placed bike stations can still play a vital role in enhancing connectivity and encouraging more sustainable forms of transportation, which can have long-term economic benefits as these areas develop.

One of the key findings of this study was the significant economic impact of locating bike stations near retail areas. Increased accessibility to bike stations tends to draw more cyclists to these areas, which, in turn, boosts local retail activities (Bencekri et al. 2024d). For example, the study found that districts such as Gangnam-gu and Seocho-gu in Seoul, both of which have a balanced land-use mix and high retail activity, were among the top-scoring areas in the suitability analysis for bike station placement (Fig. 6.32). These districts benefit from the increased foot traffic generated by bike stations, leading to higher sales for local businesses and a more vibrant street life (Bencekri et al. 2024d).

### 6.6.3 Mobility Hubs as Catalysts for Economic Growth

Mobility hubs are centralized locations where various modes of transportation such as public transit, bike-sharing, electric scooters, and ride-sharing services converge. These hubs are designed to facilitate seamless multimodal travel, making it easier for residents to switch between different forms of transportation. In the context of the 15-minute city, mobility hubs are not only critical for improving transportation efficiency but also for driving economic growth by enhancing access to businesses, services, and employment centres.

The study by Bencekri et al. (2024a) on the optimization of mobility hubs using artificial intelligence techniques (Fig. 6.6) provides a comprehensive analysis of how these hubs can be strategically placed to maximize economic and environmental benefits. This study utilized a three-tier AI-based methodology to optimize the placement of mobility hubs, focusing on key factors such as land-use mix, population and employment densities, and proximity to parking, biking, and transit facilities (Bencekri et al. 2024a).

One significant finding of this study is the substantial economic savings associated with well-placed mobility hubs. The post-assessment analysis, conducted using traffic assignment models, revealed significant reductions in vehicle kilometres travelled (VKT) and vehicle hours travelled (VHT), which translated into annual economic savings of approximately 225.9 million USD (Table 6.1). These savings are primarily attributed to reduced operational costs, travel time savings, and lower accident rates, demonstrating the economic efficiency of strategically placed mobility hubs.

Moreover, the placement of mobility hubs in high-density commercial areas, such as Yeoksam 1-dong and Myeongdong in Seoul, was found to improve connectivity to major economic centres (Bencekri et al. 2024d). The increased accessibility provided by these hubs enhances the flow of goods, services, and people, which drives economic activity. For example, businesses located near well-connected mobility hubs can attract more customers and employees, leading to higher revenue and greater overall economic vitality.

In addition to these direct economic benefits, mobility hubs contribute to broader urban development goals by promoting the use of sustainable transportation modes. By reducing the need for private vehicles and encouraging the use of public transit, cycling, and walking, mobility hubs help create more liveable and sustainable urban environments. This, in turn, makes cities more attractive to residents, businesses, and investors, which further drives economic growth.

### 6.6.4 Integrating Bikes and Mobility Hubs for Economic Gain

The full economic potential of a 15-minute city can only be realized through the strategic integration of bike stations and mobility hubs. By aligning these infrastructures, cities can create cohesive and efficient transportation networks that support local economies while promoting sustainable mobility. This integrated approach is particularly important in dense urban environments, where space is limited, and the demand for efficient transportation is high.

For example, placing bike stations near mobility hubs can significantly enhance first- and last-mile connectivity, making it easier for residents to complete their journeys without relying on private vehicles. This not only improves accessibility, but also increases foot traffic in the areas surrounding the hubs, which can boost local retail activity and support the growth of small businesses. The findings from the post-assessment analysis of mobility hubs in Seoul highlight the importance of considering both transportation efficiency and economic potential when planning such infrastructure.

Furthermore, the integration of bike stations and mobility hubs can help to create a more resilient and adaptable transportation network. By providing residents with multiple transportation options, cities can reduce their vulnerability to disruptions in any transportation mode. Resilience is particularly important in the face of challenges such as climate change, economic fluctuations, and public health crises, which can all impact the functioning of urban transportation systems.

The strategic placement of bike stations and mobility hubs also offers opportunities for public–private partnerships, which can further enhance their economic impact. For instance, cities can collaborate with private sector partners to develop and operate bike-sharing systems and mobility hubs, leveraging private investments to expand and improve these services. Such partnerships can help reduce the financial burden on city governments while ensuring that the infrastructure is well maintained and meets the needs of users.

### 6.6.5 Policy Implications and Future Directions

As cities seek to implement the 15-minute city model, it is crucial that they consider the economic potential of bike stations and mobility hubs when planning their locations. Policymakers should prioritize areas with high economic activity, such as commercial districts and transit hubs, to maximize the economic benefits of such infrastructure.

In addition, cities should adopt a flexible and adaptive approach to urban planning, allowing for ongoing evaluation and adjustment of bike stations and mobility hub placements based on changing economic and transportation patterns. This may involve the use of smart city technologies such as real-time data analytics and AI-based modelling to continuously monitor and optimize the performance of these infrastructures.

Finally, the integration of bike stations and mobility hubs should be supported by broader urban policies to promote sustainable transportation and economic development. This could include incentives for businesses located near these infrastructures, such as tax breaks or grants for sustainable development projects, as well as public awareness campaigns to encourage the use of bicycle sharing and public transit.

### 6.6.6 Conclusion

The strategic placement of bike stations and mobility hubs is essential for realizing the economic potential of a 15-minute city. By optimizing the locations of these infrastructures, cities can enhance their transportation efficiency, stimulate local economic activity, and create more resilient and sustainable urban environments. Bencekri et al. (2023b) and Bencekri et al. (2024a) provided valuable insights into how bike stations and mobility hubs can be strategically integrated into urban planning to achieve these goals. As cities continue to evolve and adapt to the challenges of the twenty-first century, the principles and methodologies outlined in these studies will be critical in guiding the development of economically prosperous and environmentally sustainable 15-Minute Cities.

## 6.7 15-Minute City for Social Equity

### 6.7.1 Introduction

Historically, urban development has been marred by significant inequalities, where access to essential services, economic opportunities, and public amenities often reflects socio-economic divisions within cities (Viana Cerqueira 2022). This disparity is particularly evident in the spatial distribution of resources, where affluent neighbourhoods typically enjoy better access to quality healthcare, education, transportation, and recreational facilities, while low-income areas are frequently underserved. The 15-minute city model, a forward-thinking urban planning concept, proposes a transformative approach to address these inequities by redesigning cities as networks of self-sufficient neighbourhoods where all residents can access essential services within a 15-minute walk or bike ride from their homes (Martin 2024). This model emphasizes the creation of sustainable, inclusive, and connected communities with a particular focus on promoting equity and social inclusion.

Equity in urban planning aims to ensure that the benefits of urban development are shared fairly among all residents, regardless of their socioeconomic status, race, ethnicity, or gender (Metzger 1996). It involves addressing systemic barriers that prevent marginalized groups from accessing the opportunities and resources they need to thrive. Social inclusion, closely tied to equity, refers to the process of improving the terms of participation in society, particularly for those who are disadvantaged through enhanced access to resources, opportunities, and services.

The 15-minute city framework provides a unique opportunity to integrate equity and social inclusion into the core of urban planning (Giuffrida et al. 2024). By decentralizing essential services and reducing the need for long commutes, the model ensures that all neighbourhoods, regardless of their economic standing, have equitable access to resources that support a high quality of life. This approach not only promotes environmental sustainability and public health but also addresses deep-rooted spatial inequalities that have long plagued urban areas.

### 6.7.2 The 15-Minute City's Equity

The 15-minute city model is designed to reconfigure urban environments to ensure equitable access to services and amenities for all residents (Allam et al. 2022a). In traditional urban settings, services such as healthcare, education, and retail are often concentrated in central or affluent areas, leaving peripheral and low-income neighbourhoods with inadequate access. This spatial inequality contributes to broader social and economic disparities, as residents in underserved areas face greater challenges in accessing jobs, education, and essential services, leading to cycles of poverty and exclusion.

In contrast, a 15-minute city seeks to create a more balanced urban landscape in which every neighbourhood is self-sufficient, with its own array of services and amenities accessible within a short distance (Giuffrida et al. 2024). This decentralized approach reduces the need for long commutes, which are often costly and time-consuming, particularly for low-income residents, who may rely on public transportation. In doing so, the 15-minute city not only reduces the financial and time burdens associated with transportation, but also enhances the quality of life by promoting walkability, reducing traffic congestion, and lowering carbon emissions.

Moreover, a 15-minute city fosters social inclusion by creating vibrant public spaces that encourage community interaction and engagement. Walkable neighbourhoods with accessible parks, community centres, and local markets become hubs of social activity where residents from diverse backgrounds can come together. This aspect of the 15-minute city is particularly important in addressing social isolation, which disproportionately affects marginalized groups including the elderly, people with disabilities, and low-income families. By creating environments that encourage social interaction, the 15-minute city can help build stronger and more cohesive communities.

However, the success of the 15-minute city in promoting equity and social inclusion hinges on the strategic placement of key infrastructure, such as mobility hubs.

The mobility hub policy is one of the key infrastructures within the 15-minute city that can significantly impact equity, and social inclusion is the mobility hub (Bencekri et al. 2023c). These hubs, where various modes of transportation such as buses, trains, bicycles, and ride-sharing services converge, are crucial in connecting residents to essential services and opportunities. When strategically located, mobility hubs can bridge gaps in transportation access, particularly in underserved communities, thereby promoting greater equity and inclusion. Mobility hubs can provide

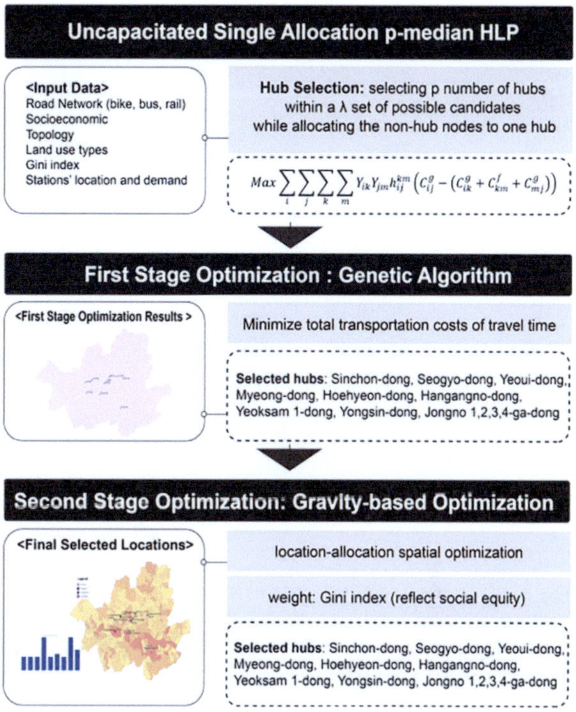


**Fig. 6.34** Equity-based allocation for mobility hubs

all residents with greater access to jobs, education, healthcare, and other essential services, thereby reducing the spatial and social barriers that contribute to social inequalities (Lopes et al. 2023).

## 6.7.3 Equity-Based Allocation of Mobility Hubs

Equitable allocation of mobility hubs is a practical application of the 15-minute city concept, which can significantly enhance social inclusion and reduce inequalities in urban settings. Bencekri et al. (2023c) offered a compelling case study on how these principles can be implemented in practice. This study focuses on Seoul, a city known for its dense population and complex social fabric, in which disparities in access to services are a critical concern.

### 6.7.3.1 Methodology and Approach

The study by Bencekri et al. (2023c) was designed to identify optimal locations for mobility hubs in Seoul that would not only enhance transportation efficiency, but also promote social equity. This study employs a multi-step methodological framework (Fig. 6.34) that incorporates traditional hub location models and advanced equity metrics, ensuring that the selected hubs would serve the needs of all socioeconomic groups.

#### Travel Time Minimization

The first stage of the study involved the use of a Genetic Algorithm (GA), a robust optimization technique, to minimize travel time across the city. GA was applied to identify potential locations for mobility hubs that would reduce the overall travel time and improve connectivity within Seoul. By focusing on areas with high transportation demand and connectivity, this study aimed to maximize the efficiency of the transportation network, ensuring that the selected hubs would serve as effective nodes within the urban system (Bencekri et al. 2023c). The emphasis on travel time minimization is crucial because shorter travel times can significantly enhance access to services for residents, particularly those in peripheral or underserved areas.

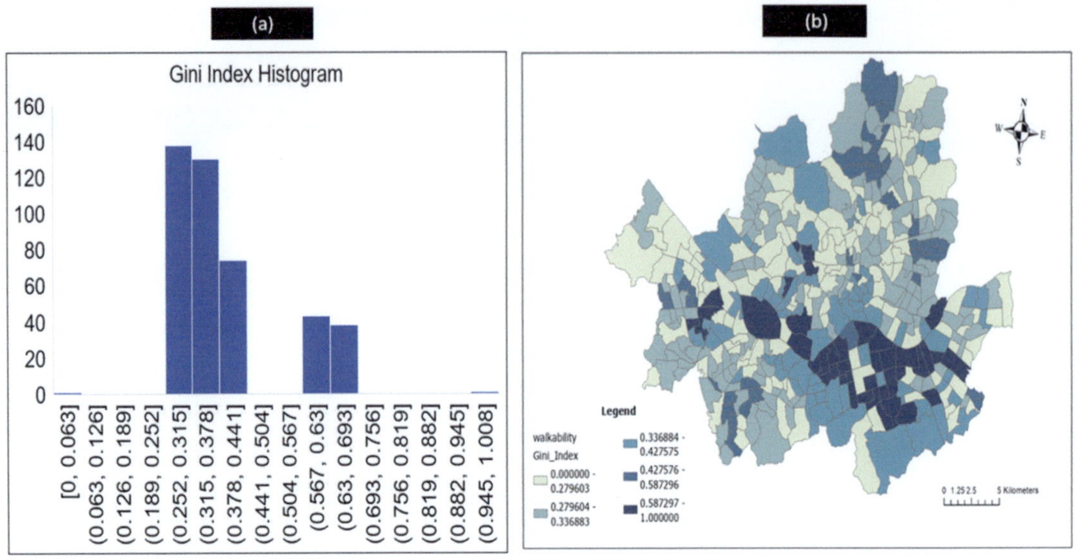


**Fig. 6.35** Gini index distribution in Seoul

### Social Equity Measurement

To integrate equity considerations into the planning process, this study employed the Gini index, a widely used measure of income inequality (Bencekri et al. 2023c). The Gini index was used to assess the distribution of income across different districts in Seoul, with higher values indicating greater income disparity (Fig. 6.35). The authors prioritized the placement of mobility hubs in areas with higher Gini index values, such as Yeouido and Jamsil 6-dong, where there is a significant need for improved transportation access among low-income residents. The use of the Gini index in this context highlights the importance of considering income inequality in urban planning to ensure that mobility hubs are in areas where they can have the most significant impact on reducing social disparities.

### Spatial Location Optimization

The final stage of the study involved a GIS-based location-allocation optimization model that incorporated the Gini index as a weighting factor (Bencekri et al. 2023c). This approach allowed the authors to identify mobility hub locations that would not only improve transportation efficiency, but also address social inequalities by enhancing access to marginalized communities. By factoring in the Gini index, this study ensured that the selected hubs would contribute to a more equitable distribution of resources and opportunities across Seoul. The use of GIS technology in this stage of the study allowed for precise analysis of spatial data, ensuring that the placement of mobility hubs was based on a comprehensive understanding of both transportation needs and social equity considerations.

## 6.7.3.2 Results and Findings

The results (Fig. 6.36) highlight the significant impact that the equity-based allocation of mobility hubs can promote social inclusion and reduce inequalities in urban environments (Bencekri et al. 2023c).

### Equity Analysis

The Gini index analysis results (Fig. 6.35) revealed significant variations in income inequality across Seoul's districts, with some areas, such as Gangnam-gu and Jongno-gu, exhibiting higher levels of disparity (Bencekri et al. 2023c). This study found that prioritizing the placement of mobility hubs in areas with higher Gini index values could play a crucial role

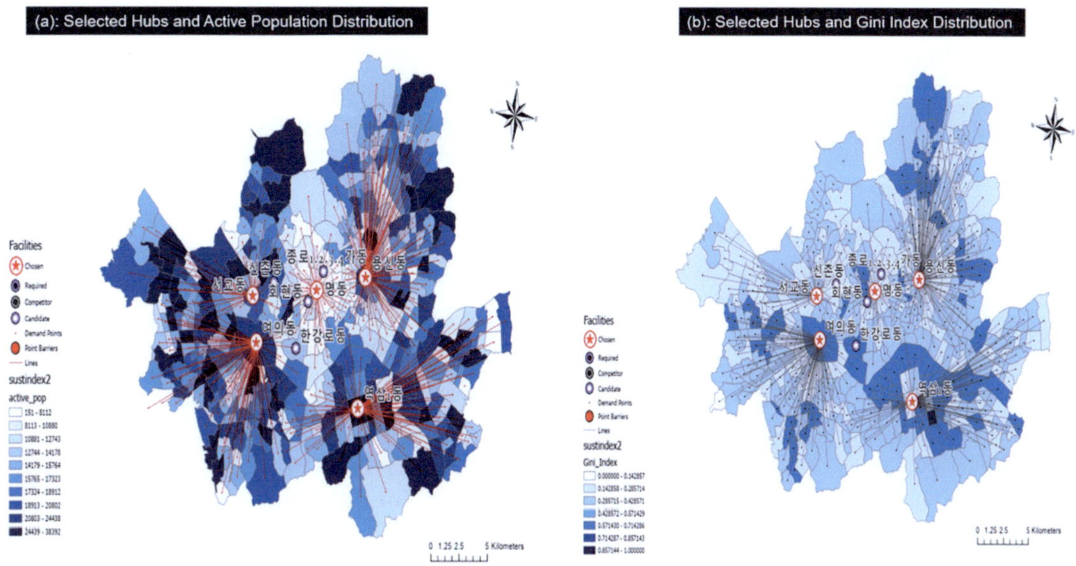


**Fig. 6.36** Selected hubs and the distribution of active population (between 15 and 65 years old) and Gini index

in bridging the transportation gap for low-income residents, thereby promoting greater equity and inclusion. By focusing on these areas, this study addressed the systemic inequalities that often leave low-income communities with limited access to essential services, exacerbating social and economic disparities (Bencekri et al. 2023c).

#### First Stage Optimization

GA-based optimization identified several key locations for mobility hubs that aligned with both transportation demand and social equity objectives (Bencekri et al. 2023c). The selected locations, including Sinchon-dong, Seogyo-dong, Yeoui-dong, Myeong-dong, Hoehyeon-dong, Hangangno-dong, Yeoksam 1-dong, Yongsin-dong, and Jongno 1,2,3,4-ga-dong, were chosen based on their potential to serve high-traffic areas and improve access for underserved populations. Each of these areas has unique characteristics that make them suitable for mobility hubs, such as a high population density, commercial activity, and cultural significance. For example, Sinchon-dong, known for its vibrant university population, was selected for its high transportation demand, whereas Yeoui-dong, a key business district, was chosen for its potential to improve access to jobs for low-income workers (Bencekri et al. 2023c).

#### Second Stage Optimization

In the second stage of optimization, the study refined the selection of mobility hubs by applying the Gini index as a weighting factor in the GIS-based location-allocation model (Bencekri et al. 2023c). The final selection included Seogyo-dong, Yeoui-dong, Myeong-dong, Yeoksam 1-dong, and Yongsin-dong (Fig. 6.36). These districts were chosen not only for their transportation efficiency but also for their potential to enhance social equity by improving access to low-income and marginalized communities. For example, Myeong-dong, a famous shopping district, was selected for its potential to enhance economic equity by improving accessibility for small business owners and market vendors, who rely on foot traffic for their livelihoods. Careful selection of these hubs reflects the study's commitment to promoting social inclusion by ensuring that all residents, regardless of their income or social status, have access to the benefits of urban mobility (Bencekri et al. 2023c).

### 6.7.4 Policy Implications

The findings of this study underscore the critical importance of equity and social inclusion in the successful implementation of the 15-minute city model. To ensure the model is both effective and inclusive, policymakers and urban planners should consider the following key recommendations:

#### 6.7.4.1 Integrate Equity Metrics into Planning

Equity must be central to the planning of mobility hubs and other urban infrastructure. While metrics like the Gini index are useful for measuring income inequality, cities should adopt additional indicators, such as access to healthcare, education, public transport, and employment opportunities. These metrics provide a more holistic understanding of social and economic disparities, enabling cities to better target underserved populations and ensure equitable access to essential services. For example, incorporating these metrics into GIS-based urban planning models can guide the equitable distribution of resources.

#### 6.7.4.2 Leverage Advanced Tools for Equity Planning

Data-driven methodologies, such as GIS-based location-allocation models and GA-based (genetic algorithm) optimization tools, offer innovative approaches to identifying optimal locations for mobility hubs that maximize both transportation efficiency and equity. Policymakers should invest in such tools to ensure urban planning decisions are evidence-based and aligned with the needs of marginalized populations. These advanced tools provide a replicable framework for other cities to strategically implement 15-minute city principles.

#### 6.7.4.3 Prioritize Infrastructure Investments in Underserved Areas

To address systemic inequalities, governments must focus on allocating resources to areas with high levels of social and economic inequality. Policies could include:

- Incentives for equity-focused projects: Tax breaks or grants for urban developers who prioritize underserved areas.
- Equitable funding allocation: Establishing criteria to ensure that infrastructure investments prioritize areas with limited access to transportation, green spaces, and essential services.

For example, Seoul's equity-based allocation of mobility hubs demonstrates how targeted investments can improve inclusivity while fostering social and economic integration.

#### 6.7.4.4 Expand Policies to Address Inequalities

While equitable mobility hubs are critical, complementary policies addressing other dimensions of inequality are necessary to create truly inclusive cities. Policymakers should simultaneously focus on:

- Expanding access to affordable housing.
- Improving the availability of quality healthcare and education.
- Promoting local employment opportunities in underserved neighbourhoods. This multidimensional approach ensures that equity is embedded across all aspects of urban planning and development, enhancing social cohesion and inclusion.

#### 6.7.4.5 Tailor Strategies to Local Contexts

The success of the 15-minute city model depends on adapting its principles to local geographic, economic, and cultural contexts. For example, cities with sprawling infrastructure or high car dependency may require phased implementation strategies, such as:

- Retrofitting existing neighbourhoods with walkable pathways and mixed-use zoning.
- Building pilot projects in selected districts to test and refine equity-focused policies before scaling citywide.

- Seoul's approach offers valuable lessons for tailoring policies to local conditions, as its focus on underserved areas successfully balanced efficiency with equity.

#### 6.7.4.6 Build Community Engagement into Planning

Community participation is essential for the successful implementation of the 15-minute city. Cities should institutionalize participatory planning processes, ensuring that residents have a voice in decision-making. Practical steps include:

- **Public consultations:** Regular workshops or forums to gather input on mobility hub locations and service improvements.
- **Citizen co-creation initiatives:** Collaborating with residents to design solutions that reflect local needs.
- **Transparency:** Transparent communication about the benefits of equitable mobility systems can also help build public support and overcome resistance to change.

#### 6.7.4.7 Strengthen Commitment to Accountability

To ensure the 15-minute city model delivers on its promises, governments must establish robust monitoring and evaluation frameworks. These systems should:

- Measure long-term outcomes, such as emissions reductions, improved accessibility, and enhanced equity.
- Use key performance indicators (KPIs) to assess progress, including metrics like proximity to services, mobility hub usage rates, and reductions in transportation-related inequalities.

Accountability mechanisms, such as regular public reporting and independent audits, can further ensure the model's success.

#### 6.7.4.8 Promote Collaboration and Innovation

Implementing the 15-minute city model requires collaboration between governments, private developers, and civil society organizations. Cities should foster partnerships that drive innovation in areas like:

- Sustainable urban technologies: For example, IoT-enabled smart mobility systems and renewable energy-powered transport hubs.
- Data integration platforms: Centralized platforms to combine demographic, geographic, and economic data for better-informed planning decisions.

Encouraging innovation ensures that cities can continually refine and optimize their implementation strategies.

### 6.7.5 Conclusion

The 15-minute city model offers a powerful framework for promoting equity and social inclusion in urban environments. By decentralizing services and integrating transportation networks that prioritize accessibility, this model can help ensure that all residents, regardless of their socioeconomic status, have equal access to the resources and opportunities they need to thrive. A case study from Seoul demonstrates how the equitable allocation of mobility hubs can play a critical role in achieving these goals. By incorporating social equity metrics into the planning process, cities can create transportation networks that are not only efficient, but also fair and inclusive, laying the foundation for more equitable and sustainable urban development.

As cities worldwide continue to grapple with the challenges of inequality and social exclusion, the principles and methods outlined in this section offer a path forward. The 15-minute city, with its focus on equity and inclusion, has the potential to transform cities, making them more just, resilient, and vibrant places to live in. By embracing this model and the lessons learned from Seoul, cities can create a future in which all residents can thrive, regardless of where they live or how much they earn.

Finally, the findings of the Seoul study underscore the importance of ongoing evaluations and adjustments in urban planning. As cities evolve and new challenges emerge, it is essential to continuously monitor the performance of mobility hubs and other infrastructure to ensure that they continue to meet the needs of all residents. This may involve revisiting hub locations, adjusting services, or incorporating new technologies to enhance accessibility and efficiency. Policymakers should also be open to feedback from residents and stakeholders using this input to refine and improve urban planning strategies over time.

## References


Allam Z, Bibri SE, Chabaud D, Moreno C (2022a) The '15-Minute City' concept can shape a net-zero urban future. Humanit Soc Sci Commun 9:126. https://doi.org/10.1057/s41599-022-01145-0

Allam Z, Bibri SE, Chabaud D, Moreno C (2022b) The theoretical, practical, and technological foundations of the 15-minute city model: proximity and its environmental, social and economic benefits for sustainability. Energies 15:6042. https://doi.org/10.3390/en15166042

Arnold T, Dale S, Timmis A et al (2023) An exploratory study of Mobility Hub implementation. Res Transp Econ 101:101338. https://doi.org/10.1016/j.retrec.2023.101338

Barbieri L, D'Autilia R, Marrone P, Montella I (2023) Graph representation of the 15-minute city: a comparison between Rome, London, and Paris. Sustainability 15:3772. https://doi.org/10.3390/su15043772

Bencekri M, Lee D, Kim S et al (2023bc) Equity-based allocation of mobility hubs in seoul. Chem Eng Trans 106:265–270. https://doi.org/10.3303/CET23106045

Bencekri M, Kim S, Van FY, Lee S (2024a) Locating carbon neutral mobility hubs using artificial intelligence techniques. Sci Rep 14:12328. https://doi.org/10.1038/s41598-024-62701-z

Bencekri M, Kim S, Van FY, Lee S (2024b) Locating carbon neutral mobility hubs using artificial intelligence techniques. Sci Rep 14:12328. https://doi.org/10.1038/s41598-024-62701-z

Bencekri M, Van Fan Y, Lee D et al (2024cd) Optimizing shared bike systems for economic gain: integrating land use and retail. J Transp Geogr 118:103920. https://doi.org/10.1016/j.jtrangeo.2024.103920

Bencekri M, Ku D, Lee D, Lee S (2023b) A planning support system for boosting bikeability in seoul. In: Intelligence for future cities. pp 149–171

Bencekri M, Lee D, Ku D, Lee S (2024c) A planning support system for boosting walkability. Proc Inst Civ Eng - Munic Eng 1–14. https://doi.org/10.1680/jmuen.23.00040

Bencekri M, Kwak J, Lee D, et al (2025) A systematic review of the 15-minute city framework: implications for environmental heritage preservation in the Anthropocene. Int J Urban Sci 1–30. https://doi.org/10.1080/12265934.2024.2445240

Casarin G, MacLeavy J, Manley D (2023) Rethinking urban utopianism: The fallacy of social mix in the 15-minute city. Urban Stud 60:3167–3186. https://doi.org/10.1177/00420980231169174

Dilks R (2021) Why mobility hubs are crucial to making transport more sustainable

Gehrke SR, Clifton KJ (2019) An activity-related land use mix construct and its connection to pedestrian travel. Environ Plan B Urban Anal City Sci 46:9–26. https://doi.org/10.1177/2399808317690157

Giuffrida N, Mölter A, Pilla F et al (2024) On the equity of the x-minute city from the perspective of walkability. Transp Eng 16:100244. https://doi.org/10.1016/j.treng.2024.100244

Jin T, Wang K, Xin Y, Shi J, Hong Y, Witlox F (2024) Is a 15-minute city within reach? Measuring multimodal accessibility and carbon footprint in 12 major American cities. Land Use Policy 142:107180. ISSN 0264-8377. https://doi.org/10.1016/j.landusepol.2024.107180

Kim Y, Choi B, Choi M, et al (2024) Enhancing pedestrian perceived safety through walking environment modification considering traffic and walking infrastructure. Front Public Heal 11. https://doi.org/10.3389/fpubh.2023.1326468

Kondratenko A (2023) 15-Minute City: Revolutionizing Urban Transportation to Combat Climate Change. https://www.digitalbluefoam.com/post/15-minute-city-revolutionizing-urban-transportation-to-combat-climate-change

Krishnan AR (2024) Research trends in criteria importance through intercriteria correlation (CRITIC) method: a visual analysis of bibliographic data using the Tableau software. Inf Discov Deliv. https://doi.org/10.1108/IDD-02-2024-0030

Lima FT, Brown NC, Duarte JP (2022) A Grammar-based optimization approach for designing urban fabrics and locating amenities for 15-minute cities. Buildings 12:1157. https://doi.org/10.3390/buildings12081157

Lopes AS, Orozco-Fontalvo M, Moura F, Vale D (2023) Mobility as a service and socio-territorial inequalities: a systematic literature review. J Transp Land Use 16:215–240. https://doi.org/10.5198/jtlu.2023.2273

Marchigiani E, Bonfantini B (2022) urban transition and the return of neighbourhood planning. questioning the proximity syndrome and the 15-minute city. Sustainability 14:5468. https://doi.org/10.3390/su14095468

Martin RJ (2024) Operationalising the 15-minute city with New Mobility services. Proc from Annu Transp Conf Aalborg Univ 31. https://doi.org/10.54337/ojs.td.v31i.8706

Metzger JT (1996) The theory and practice of equity planning: an annotated bibliography. J Plan Lit 11:112–126. https://doi.org/10.1177/088541229601100106

Mishra MDU, Kumar MS, Das MS (2024) Achieving accessibility through promoting non-motorized transport within the neighborhood level of city: a case study of ward 25 & 27, Bhubaneswar. Int J Res Appl Sci Eng Technol 12:4677–4682. https://doi.org/10.22214/ijraset.2024.62636

Moreno C, Allam Z, Chabaud D et al (2021) Introducing the "15-Minute City": sustainability, resilience and place identity in future post-pandemic cities. Smart Cities 4:93–111. https://doi.org/10.3390/smartcities4010006

Moreno C, Thorne M, Gehl J (2024) The 15-minute city: a solution to saving our time and our planet

Tran M, Draeger C (2021) A data-driven complex network approach for planning sustainable and inclusive urban mobility hubs and services. Environ Plan B Urban Anal City Sci 48:2726–2742. https://doi.org/10.1177/2399808320987093

Venerandi A, Mellen H, Romice O, Porta S (2024) Walkability indices—The state of the art and future directions: a systematic review. Sustainability 16:6730. https://doi.org/10.3390/su16166730

Viana Cerqueira ED (2022) Inequalities in access to urban services. in: cities at the heart of inequalities. Wiley, pp 79–103

Wey W-M, Chiu Y-H (2013) Assessing the walkability of pedestrian environment under the transit-oriented development. Habitat Int 38:106–118. https://doi.org/10.1016/j.habitatint.2012.05.004

Zong H (2023) Discussion on "15-Minute City" from the perspective of neighborhood effect and urban facilities. Lect Notes Educ Psychol Public Media 10:249–255. https://doi.org/10.54254/2753-7048/10/20230224

# Guideline for a 15-Minute City

# 7

## 7.1 Introduction

The 15-minute city represents a transformative vision for urban living, reimagining cities as networks of interconnected neighbourhoods where essential services, such as healthcare, education, employment, and recreation, are accessible within a 15-minute walk or bike ride. By prioritizing accessibility, sustainability, and community well-being, this model aims to create cities that are not only functional but also inclusive, vibrant, and resilient in the face of modern challenges.

Key principles of the 15-minute city include integrating spatial and temporal dimensions, leveraging smart technologies, and engaging communities in urban planning. Effective city design requires optimizing the physical layout of urban areas while also considering the timing of activities, ensuring that services are accessible throughout the day and adaptable to evolving resident needs. Smart technologies, such as real-time monitoring, predictive analytics, and IoT-enabled systems, enhance urban efficiency and sustainability by optimizing mobility, reducing emissions, and improving resource management. Digital platforms for citizen engagement also empower residents to participate actively in decision-making, fostering a sense of ownership and alignment between urban policies and community priorities.

The 15-minute city supports environmental sustainability by reducing car dependency, encouraging active transportation (walking and cycling), and expanding green infrastructure. These strategies not only lower greenhouse gas emissions but also improve air quality and promote healthier lifestyles. Furthermore, the model strengthens urban resilience by creating adaptable, multi-use spaces that can recover from environmental, social, and economic disruptions.

At the core of the 15-minute city is a commitment to equity and inclusion. To ensure that all residents, particularly marginalized and underserved populations, benefit from this model, urban planning must focus on fair access to services and amenities, affordable housing, and mobility solutions. Policies should address systemic inequalities by prioritizing investments in underserved neighbourhoods and tailoring interventions to local contexts. By embedding equity as a guiding principle, the 15-minute city can create truly inclusive environments where all residents, regardless of their socioeconomic status, can thrive.

The 15-minute city offers a framework for rethinking urban environments to balance functionality, sustainability, and inclusivity. By integrating smart technologies, fostering community engagement, and ensuring equitable access to resources, this model has the potential to transform cities into resilient, liveable spaces for all.



M. Bencekri and S. Lee, *Sustainable Urban Evolution*, Sustainable Development Goals Series,
https://doi.org/10.1007/978-981-96-5146-7_7

## 7.2 Opportunities and Challenges

The 15-minute city model offers a transformative vision for urban planning, focusing on sustainability, inclusivity, and resilience by ensuring that essential services are accessible within a short walk or bike ride from residents' homes. This innovative approach addresses many of the pressing challenges of contemporary cities, including car dependency, environmental degradation, and economic inequality, while fostering vibrant local communities. However, realizing the full potential of the 15-minute city requires overcoming structural, social, and financial obstacles.

The model provides significant opportunities to improve urban **liveability**. By bringing essential services, within close proximity, the 15-minute city reduces commuting times and allows residents to spend more time on personal and community activities. Shorter commutes have been shown to increase life satisfaction and lower stress, improving mental well-being. Additionally, local living strengthens **community bonds** as residents engage with neighbours and participate in shared initiatives. This creates neighbourhoods that are not only convenient but also socially connected and supportive, enhancing overall quality of life.

Environmentally, the 15-minute city contributes to global **sustainability** goals by reducing car dependency and promoting active transportation, such as walking and cycling. Fewer cars on the road result in lower greenhouse gas emissions and improved air quality, directly benefiting public health. The integration of green spaces, such as parks, urban gardens, and tree-lined streets, further enhances biodiversity, provides natural cooling, and strengthens climate resilience. Cities like Paris have demonstrated how such approaches can lead to significant reductions in carbon emissions while creating healthier, more sustainable urban environments.

The model also supports **economic growth** by revitalizing local economies. Walkable, mixed-use neighbourhoods encourage residents to shop locally, increasing foot traffic for small businesses and stimulating community-based economic activity. These neighbourhoods attract investment and create jobs while enhancing urban resilience by reducing reliance on centralized commercial hubs. For example, studies show that walkable and bikeable areas often enjoy higher property values and stronger economic performance, underscoring the long-term financial benefits of localized urban living.

Despite these opportunities, the implementation of the 15-minute city faces several challenges. One significant barrier is **outdated zoning laws and building codes**, which often restrict the development of mixed-use, walkable neighbourhoods. Traditional single-use zoning separates residential, commercial, and industrial areas, limiting the flexibility required for the 15-minute city's integrated design. **Reforming these regulations** is essential to enable higher-density, mixed-use developments and adaptive reuse of existing structures. Cities like Paris and Melbourne have successfully modernized their zoning codes to prioritize pedestrian-friendly neighbourhoods, offering a replicable model for other urban areas.

Another critical challenge is the risk of **gentrification and displacement**. Revitalizing neighbourhoods to align with 15-minute city principles can lead to rising property values, increased rents, and the displacement of long-term, lower-income residents. To mitigate this, policymakers must implement **inclusive housing policies**, such as rent control, subsidies for affordable housing, and mandatory allocations for low- and moderate-income families in new developments. **Community land trusts** (CLTs) and **cooperative housing models** can also ensure long-term affordability and prevent speculative property price increases. Boston's success with CLTs demonstrates how community ownership can preserve affordability while fostering equitable growth.

**Securing adequate funding** is another hurdle. Large-scale urban transformations require substantial investments in infrastructure, transportation, and public services. Traditional funding mechanisms often fall short, necessitating innovative financing solutions. **Value-capture financing**, where the increase in property values generated by new developments helps fund infrastructure upgrades, offers one such approach. **Tax increment financing (TIF) and municipal bonds** are additional tools to secure funding. **Public–private partnerships (PPPs)** can also play a critical role by leveraging private investment while maintaining public oversight to ensure equity and accountability. Bogotá's transit expansion illustrates how PPPs can deliver essential infrastructure improvements effectively.

Finally, integrating **smart technologies** into the 15-minute city framework presents both opportunities and challenges. While technologies like IoT devices, digital twins, and predictive analytics can enhance urban mobility, monitor environmental impacts, and optimize resource use, their implementation requires significant **investments and technical expertise**. Additionally, the increased reliance on digital systems raises concerns about **data privacy and the digital divide**. **Robust data protection policies, affordable internet access, and digital literacy programs** are essential to ensure that all residents, particularly marginalized groups, can benefit from these advancements. Singapore's "Smart Nation" initiative offers a blueprint for integrating technology equitably and securely.

To address these challenges, cities must adopt an integrated, forward-looking approach to urban planning. For example, zoning reforms can not only enable mixed-use developments but also mandate affordable housing to mitigate displacement risks. Similarly, value-capture financing can fund both infrastructure projects and inclusive housing initiatives, creating synergies between economic and social goals. Policymakers must ensure that solutions are adaptable to diverse urban contexts, allowing the 15-minute city model to scale effectively while addressing future challenges, such as climate change and demographic shifts.

In conclusion, the 15-minute city offers a visionary framework for building sustainable, inclusive, and resilient urban environments. However, its success hinges on overcoming regulatory, financial, and technological barriers while safeguarding equity and inclusivity. With strategic planning, innovative policies, and an integrated approach, cities can unlock the full potential of the 15-minute city, paving the way for a sustainable future.

## 7.3 Guideline for a 15-Minute City

The 15-minute city presents a transformative vision for urban planning, aiming to create sustainable, inclusive, and resilient neighbourhoods where essential services are accessible within a short walk or bike ride. This guideline offers a practical framework for **urban planners, policymakers, and stakeholders** to implement the model, integrating insights from previous chapters and addressing the need for context-sensitive solutions. By combining policy reform, infrastructure investment, urban design, smart technologies, equity measures, and environmental strategies, cities can realize the full potential of the 15-minute city while addressing modern urban challenges.

Recognizing that cities vary widely in terms of geography, demographics, and existing infrastructure, this guideline (Fig. 7.1) emphasizes the need for flexible, adaptable strategies that respond to local contexts. The following sections outline a roadmap for implementation, linking each component of the 15-minute city model into a cohesive and actionable plan.

### 7.3.1 Reforming Policies and Governance

Achieving the 15-minute city vision requires fundamental policy reforms to enable mixed-use development, encourage sustainable practices, and foster inclusive governance structures.

#### 7.3.1.1 Zoning and Land-Use Reform

Traditional zoning laws, which segregate residential, commercial, and industrial areas, hinder the creation of walkable, mixed-use neighbourhoods. Cities must transition to flexible zoning codes that allow for higher-density, mixed-use

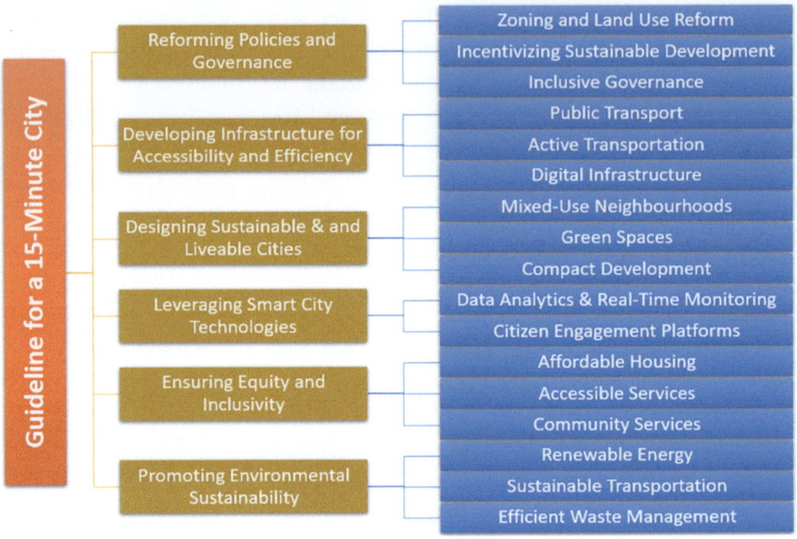


**Fig. 7.1** Guideline for a 15-minute city

developments. Adaptive reuse of existing structures should also be prioritized to reduce environmental impact, preserve cultural heritage, and maximize resource efficiency. For example, cities like Portland and Vancouver have successfully revised zoning regulations to promote compact, sustainable urban growth.

#### 7.3.1.2 Incentivizing Sustainable Development

Financial incentives are essential for encouraging green building practices, renewable energy adoption, and transit-oriented developments. Cities can provide subsidies, tax breaks, and grants to developers and residents who adopt environmentally friendly practices, such as solar panel installation or energy-efficient construction. Successful examples include Copenhagen's renewable energy incentives and Freiburg's transit-focused urban planning.

#### 7.3.1.3 Inclusive Governance

Robust community engagement is central to implementing the 15-minute city. Cities should establish advisory boards, conduct public consultations, and leverage digital platforms to ensure diverse voices are included in the planning process. Participatory planning in Barcelona and Melbourne has demonstrated that involving residents fosters a sense of ownership and produces outcomes that better reflect community needs.

### 7.3.2 Developing Infrastructure for Accessibility and Efficiency

Infrastructure development is crucial to enabling the accessibility and connectivity central to the 15-minute city model. Investments should prioritize public transport, active transportation networks, and digital infrastructure.

#### 7.3.2.1 Public Transport

A reliable, affordable, and integrated public transit system is foundational to the 15-minute city. Expanding networks to underserved areas and ensuring seamless connectivity between buses, trams, and trains can reduce car dependency and improve urban mobility. Real-time tracking systems and smart ticketing can enhance convenience and reliability. Tokyo's extensive rail network and Zurich's integrated public transport system serve as models for efficient urban transit.

#### 7.3.2.2 Active Transportation

Promoting walking and cycling requires investments in pedestrian-friendly streets, dedicated bike lanes, and related infrastructure like bike-sharing programs and secure parking. Amsterdam and Copenhagen illustrate how prioritizing active transportation reduces congestion, improves public health, and fosters vibrant neighbourhoods.

#### 7.3.2.3 Digital Infrastructure

Robust digital infrastructure enables smart city applications that optimize urban living, such as traffic management, energy efficiency, and real-time public service updates. High-speed internet is essential for remote work, online education, and telehealth services. Singapore's "Smart Nation" initiative highlights how IoT-enabled systems can enhance urban operations and improve residents' quality of life.

### 7.3.3 Designing a Sustainable and Liveable Cities

Effective urban design ensures that neighbourhoods are compact, vibrant, and adaptable to residents' needs. Key strategies include mixed-use development, integration of green spaces, and compact urban forms.

#### 7.3.3.1 Mixed-Use Neighbourhoods

Designing neighbourhoods where residential, commercial, and recreational spaces coexist reduces commuting times, promotes local businesses, and fosters community interaction. Examples like New York's Greenwich Village and Tokyo's Shibuya district showcase the economic and social benefits of mixed-use developments.

#### 7.3.3.2 Green Spaces

Accessible parks, urban gardens, and tree-lined streets provide numerous environmental and social benefits, including reducing heat islands, improving air quality, and enhancing mental well-being. Singapore and Vancouver demonstrate how integrating green spaces into urban planning creates more sustainable and liveable environments.

#### 7.3.3.3 Compact Development

Encouraging higher-density development reduces urban sprawl, conserves land and ensures that essential services remain within close reach. Hong Kong and Barcelona exemplify how compact urban designs can support vibrant, efficient, and sustainable city living.

### 7.3.4 Leveraging Smart City Technologies

Advanced technologies play a critical role in implementing the 15-minute city by enabling data-driven decision-making and responsive urban systems.

#### 7.3.4.1 Data Analytics and Real-Time Monitoring

Cities should utilize predictive analytics and IoT-enabled systems to monitor traffic, energy use, and public services in real time. This data allows planners to anticipate challenges and allocate resources proactively. Examples from London and Singapore show how smart technologies improve urban efficiency and resilience.

#### 7.3.4.2 Citizen Engagement Platforms

Digital platforms facilitate two-way communication between residents and planners, ensuring that urban development reflects community needs. Tools like FixMyStreet in the UK and Citizen Lab empower residents to report issues, provide feedback, and participate in planning.

### 7.3.5 Ensuring Equity and Inclusivity

Equity is a cornerstone of the 15-minute city, ensuring all residents, regardless of income or ability, benefit from accessible services and opportunities.

#### 7.3.5.1 Affordable Housing

Policies such as rent control, housing subsidies, and community land trusts prevent displacement and ensure diverse communities. Vienna and Singapore exemplify successful integration of affordable housing into urban planning.

#### 7.3.5.2 Accessible Services

Urban design must prioritize accessibility for all, including individuals with disabilities. Features such as wide sidewalks, ramps, and inclusive public spaces ensure equal opportunities for mobility and recreation.

#### 7.3.5.3 Community Services

Investments in local schools, healthcare facilities, and social services are essential for building cohesive and self-sufficient neighbourhoods. Toronto and Melbourne provide strong examples of how robust community infrastructure enhances quality of life.

### 7.3.6 Promoting Environmental Sustainability

Environmental stewardship is integral to the 15-minute city, with strategies focused on renewable energy, sustainable transportation, and waste management.

#### 7.3.6.1 Renewable Energy

Incentives for solar panels, wind turbines, and geothermal systems reduce reliance on fossil fuels. Freiburg and San Francisco demonstrate how renewable energy integration can lower emissions and operational costs.

#### 7.3.6.2 Sustainable Transportation

Electric vehicle infrastructure, car-sharing programs, and efficient public transit reduce emissions and support clean urban mobility. Oslo and Amsterdam offer exemplary practices in sustainable transport systems.

#### 7.3.6.3 Efficient Waste Management

Advanced waste collection systems, recycling programs, and circular economy initiatives minimize landfill use and conserve resources. San Francisco and Sweden's waste-to-energy initiatives showcase best practices in sustainable waste management.

## 7.4 Guideline's Implementation Roadmap

The 15-minute city model requires a structured and adaptive roadmap to guide cities in transitioning to vibrant, sustainable, and inclusive urban environments. This roadmap provides a phased implementation process, balancing community needs, technological advancements, and local contexts. By following these steps, cities can address urban challenges such as accessibility, sustainability, and equity while building resilient neighbourhoods that thrive for generations.

Recognizing that cities differ in their infrastructure, population density, and socioeconomic conditions, this roadmap (Fig. 7.2) is designed to be flexible and responsive to diverse urban contexts. Each phase builds on the previous one, creating an iterative process of learning, scaling, and refining.

### 7.4.1 Assessment: Establishing the Baseline

The first step in implementing the 15-minute city is conducting a thorough assessment to identify gaps in infrastructure, service accessibility, and urban planning policies.

- Key Actions:
  - Map existing services, such as schools, healthcare, and retail, to determine accessibility gaps.

**Fig. 7.2** Roadmap implementation for the guideline

- Audit transportation networks, including walkability, cycling paths, and public transit coverage, to identify underserved areas.
- Review zoning laws and land use policies to assess whether they allow for mixed-use, high-density development.
- Conduct stakeholder consultations with local communities, businesses, and urban planners to understand needs and priorities.

- Purpose: To create a data-driven foundation for decision-making, ensuring that interventions are targeted and relevant.
- Considerations for Context: In cities with limited resources, prioritize underserved neighbourhoods and areas with high socioeconomic disparities.

### 7.4.2 Pilot Projects: Testing and Learning

Pilot projects serve as small-scale, controlled experiments to test the feasibility and impact of the 15-minute city principles before citywide implementation.

- Key Actions:
  - Select diverse pilot neighbourhoods that represent different socioeconomic, cultural, and geographic conditions.
  - Implement mixed-use zoning reforms, pedestrian-friendly infrastructure, bike lanes, and public transport improvements in these areas.
  - Introduce smart technologies such as IoT sensors, real-time public transport tracking, and participatory planning platforms to enhance urban efficiency and gather feedback.
  - Offer financial incentives for local businesses, green building practices, and renewable energy adoption.
- Purpose: To evaluate the effectiveness of proposed strategies, gather feedback from residents, and refine implementation plans.
- Example: Paris's Quartier du Grand Parc successfully piloted 15-minute city principles, demonstrating the viability of mixed-use, walkable neighbourhoods with strong community engagement.

### 7.4.3 Community Engagement: Building Trust and Ownership

The active involvement of residents and stakeholders is essential throughout the implementation process to ensure alignment with community needs and foster long-term support.

- Key Actions:
  - Organize town halls, focus groups, and online forums to gather feedback on pilot projects and proposed initiatives.
  - Use participatory planning tools, such as mobile apps and surveys, to involve residents in decision-making.
  - Create advisory boards composed of diverse community representatives to provide ongoing input and ensure inclusivity.
- Purpose: To foster transparency, build trust, and create a sense of shared ownership among residents, ensuring that the 15-minute city reflects local aspirations.
- Example: Barcelona's participatory planning process for *Superblocks* demonstrates how active resident involvement can lead to sustainable and community-centred urban transformations.

### 7.4.4 Scaling up: Expanding Successful Initiatives

Once pilot projects are refined, successful strategies can be expanded citywide while tailoring solutions to local conditions.

- Key Actions:
  - Implement zoning reforms and infrastructure upgrades across multiple neighbourhoods, prioritizing areas with the greatest need or opportunity for improvement.
  - Integrate pilot-tested active transportation networks, public transit enhancements, and digital tools into broader urban planning strategies.
  - Collaborate with regional governments, private developers, and civic organizations to coordinate large-scale implementation efforts.
- Purpose: To create a network of interconnected 15-minute neighbourhoods, ensuring equitable access to services citywide.
- Considerations for Context: In dense urban areas, focus on retrofitting existing neighbourhoods; in suburban or sprawling areas, prioritize compact development and public transit expansion.

### 7.4.5 Monitoring and Evaluation: Ensuring Long-Term Success

Continuous monitoring and evaluation are critical for tracking progress, refining strategies, and maintaining the momentum of 15-minute city initiatives.

- Key Actions:
  - Establish clear performance metrics, such as reductions in commute times, increases in active transportation use, or decreases in car dependency and emissions.
  - Use real-time data from IoT systems to track urban mobility, energy efficiency, and service accessibility.
  - Regularly solicit feedback from residents through surveys, public meetings, and digital platforms to assess satisfaction and identify emerging challenges.
  - Share progress transparently with stakeholders and residents to maintain accountability.
- Purpose: To measure outcomes, ensure adaptability, and demonstrate the long-term value of the 15-minute city model.
- Example Metrics: Percentage of residents within a 15-minute walk of services, improvements in air quality, or increases in local business revenue.

### 7.4.6 Long-Term Vision and Flexibility

To ensure the roadmap remains effective over time, cities must:

- Adapt to Demographic Changes: Account for population growth, aging populations, and migration patterns to maintain accessibility for all residents.
- Leverage Emerging Technologies: Stay updated on advancements in urban tech, such as AI-driven urban analytics or green construction materials, to refine strategies.
- Institutionalize the Model: Embed 15-minute city principles into long-term urban governance frameworks to sustain progress across political cycles.

## 7.5 Conclusion

The 15-minute city offers a transformative vision for urban living, emphasizing accessibility, sustainability, and equity to create inclusive and resilient communities. By ensuring that essential services are accessible within walking or biking distance, this model addresses pressing urban challenges such as car dependency, environmental degradation, and socioeconomic inequality. Through a holistic framework, integrating policy reform, infrastructure investment, smart technologies, and inclusive urban design, cities can reimagine their spaces as hubs of opportunity and well-being for all residents.

This chapter highlights the critical opportunities presented by the 15-minute city, including reduced commuting times, healthier lifestyles, strengthened local economies, and improved environmental outcomes. Successful examples from around the world demonstrate how flexible zoning, green infrastructure, and equitable mobility solutions can advance these goals. However, challenges such as outdated zoning laws, the risk of gentrification, funding constraints, and the digital divide must be addressed to ensure successful implementation.

The roadmap outlined in this chapter provides a practical and adaptable strategy for cities to adopt the 15-minute city model. Key steps include reforming zoning laws to enable mixed-use developments, investing in public and active transportation networks, leveraging smart technologies for real-time monitoring and citizen engagement, and prioritizing equity through affordable housing and accessible services. Continuous monitoring and community participation will be essential to refine strategies and meet evolving urban needs.

By embracing the principles of the 15-minute city and tailoring solutions to local contexts, policymakers and urban planners can build more just, sustainable, and connected cities. This model represents not just an urban planning framework but a blueprint for a future where cities are designed to thrive in the face of modern challenges, fostering vibrant, equitable, and sustainable communities for generations to come.

## Uncited References


Al-Raeei M (2024) The smart future for sustainable development: artificial intelligence solutions for sustainable urbanization. Sustain Dev. https://doi.org/10.1002/sd.3131

Dąbrowska-Żółtak K, Wojtowicz J, Wrona S (2021) Reconfigurable neighborhood—mechatronisation of the urban design. Sustainability 13:13547. https://doi.org/10.3390/su132413547

Gil-Rivas V, Kilmer RP (2016) Building community capacity and fostering disaster resilience. J Clin Psychol 72:1318–1332. https://doi.org/10.1002/jclp.22281

Jin T, Wang K, Xin Y et al (2024) Is a 15-minute city within reach? Measuring multimodal accessibility and carbon footprint in 12 major American cities. Land Use Policy 142:107180. https://doi.org/10.1016/j.landusepol.2024.107180

Koch M, Grubb M (2023) Agenda 21. The earthscan reader in sustainable development. Routledge, London, pp 191–193

Ligmann-Zielinska A, Church RL, Jankowski P (2008) Spatial optimization as a generative technique for sustainable multiobjective land-use allocation. Int J Geogr Inf Sci 22:601–622. https://doi.org/10.1080/13658810701587495

Munn K, Dragićević S (2023) Towards a spatio-temporal multicriteria evaluation method: a suitability analysis of residential units in a 3D urban environment. Trans GIS 27:1830–1845. https://doi.org/10.1111/tgis.13101

Nair R, Miller-Hooks E, Hampshire RC, Bušić A (2013) Large-scale vehicle sharing systems: analysis of Vélib'. Int J Sustain Transp 7:85–106. https://doi.org/10.1080/15568318.2012.660115

Nemeškal J, Ouředníček M, Pospíšilová L (2020) Temporality of urban space: daily rhythms of a typical week day in the Prague metropolitan area. J Maps 16:30–39. https://doi.org/10.1080/17445647.2019.1709577

Nižetić S, Djilali N, Papadopoulos A, Rodrigues JJPC (2019) Smart technologies for promotion of energy efficiency, utilization of sustainable resources and waste management. J Clean Prod 231:565–591. https://doi.org/10.1016/j.jclepro.2019.04.397

Taylor AA, Khan MY (2018) Federal air and water regulations. In: Regulatory toxicology, 3rd edition. CRC Press, Boca Raton, Florida [2019], pp 265–280

Williams K (2014) Urban form and infrastructure : a morphological review. Future cities working paper 58

Yang J (2023) Visualizing and assessing the 15-minute city facility configuration of city region A study on the Beijing-Tianjin-Hebei Urban Agglomeration. Adv Educ Humanit Soc Sci Res 4:63. https://doi.org/10.56028/aehssr.4.1.63.2023

MIX
Papier aus verantwortungsvollen Quellen
Paper from responsible sources
FSC® C105338

If you have any concerns about our products,
you can contact us on
**ProductSafety@springernature.com**

In case Publisher is established outside the EU,
the EU authorized representative is:
**Springer Nature Customer Service Center GmbH**
**Europaplatz 3, 69115 Heidelberg, Germany**

Printed by Libri Plureos GmbH
in Hamburg, Germany